Medical Device Design
Innovation from Concept to Market

Medical Device Design

Innovation from Concept to Market

Peter J. Ogrodnik

AMSTERDAM • BOSTON • HEIDELBERG • LONDON
NEW YORK • OXFORD • PARIS • SAN DIEGO • SAN FRANCISCO
SINGAPORE • SYDNEY • TOKYO

Academic Press is an imprint of Elsevier

Academic Press is an imprint of Elsevier
The Boulevard, Langford Lane, Kidlington, Oxford, OX5 1GB, UK
84 Theobald's Road, London WC1X 8RR, UK

First edition 2013

Notices
Knowledge and best practice in this field are constantly changing. As new research and experience
broaden our understanding, changes in research methods, professional practices, or medical treatment
may become necessary.

Practitioners and researchers must always rely on their own experience and knowledge in evaluating
and using any information, methods, compounds, or experiments described herein. In using such
information or methods they should be mindful of their own safety and the safety of others, including
parties for whom they have a professional responsibility.

To the fullest extent of the law, neither the Publisher nor the authors, contributors, or editors, assume
any liability for any injury and/or damage to persons or property as a matter of products liability,
negligence or otherwise, or from any use or operation of any methods, products, instructions, or ideas
contained in the material herein.

British Library Cataloguing in Publication Data
A catalogue record for this book is available from the British Library

Library of Congress Cataloging-in-Publication Data
A catalog record for this book is available from the Library of Congress

ISBN: 978-0-12-391942-7

For information on all Academic Press publications
visit our website at elsevier.com

Printed and bound in the United States of America

12 10 9 8 7 6 5 4 3 2 1

Working together to grow
libraries in developing countries

www.elsevier.com | www.bookaid.org | www.sabre.org

ELSEVIER   BOOK AID
International   Sabre Foundation

I would like to dedicate this book to the three most important people in my life — my wife Lynda and my two daughters Natasha and Zoë.

Beware of raining sheep!

Contents

Preface

This book fulfills an ambition I have had for many years. When I first started on the medical engineering pathway I was so disappointed that there were no texts to help me design a device – Mrs. Hubbard's bookshelves were empty. There were loads of books available to tell me how to measure the angle of an ankle joint, or even how to use an x-ray machine…but could I find one telling me how to design one? No. Luckily my background is in mechanical engineering and good design practice was forced into me so adhering to medical device regulations was relatively easy. However, I was dismayed when I looked back and saw how much time I had wasted trying to make the basic design principles fit into the regulatory framework.

About a year ago I looked on the bookshelves again (only now Mrs. Hubbard's cupboard was electronic and web based). Once again I was dismayed that little progress had been made. The regulatory bodies themselves had come up with some guidelines, but all of the biomedical engineering books had followed the same old path. It was at this point that my ambition was rekindled and I decided to contact the publisher of my first book to see if they were interested in having a medical devices design handbook in their portfolio. Unfortunately it was out of their scope but they gave me a name of someone they knew who might be interested. I therefore, with some trepidation, sent a brief proposal to this person. Little known to me was that at that very moment, he was in a meeting in the USA discussing the need for a medical devices design handbook with his colleagues, and they were trying to think who could write it; his email went ping and there was my proposal. Serendipity does throw up some unlikely coincidences – and this was one of the best.

This was not my first serendipitous event. I only came into medical engineering because of answering a telephone in an office I just happened to be passing – on a day that I happened to be at work during summer vacation – and talking to a then complete stranger (who became my friend of 21 years, Peter T) about something called a tibia. Several commercial products, numerous patents, and papers galore followed.

Back to the story. After a frantic email exchange I set about designing THE medical devices design handbook, only to realize that I was probably doing the same as others before me, that is, ignoring the fact that this is about design in a medical devices environment, not just biomedical engineering. I soon came to realize that the best approach would be to start from the beginning (in design terms) and assume all readers (no matter what background) have a poor design education, and to take things step by step. Once I had identified this simple seed of an idea, the book fell into place. In fact, I had actually designed the book not just written it. I decided that the format for the text must be informal; it must have the feel of a conversation rather than the usual, dry – sometimes pompous – nature of textbooks. I hope I have achieved this (not the pompous bit!). The main layout of the book is bite-sized, self-contained chapters that you can read or not read as you choose. So, for example, if you are happy with your knowledge of labeling there is little need for you to spend hours reading that chapter…but you may spend more time on specification development. Oh, while we are on that subject, DO NOT ignore the specification chapter – this is the core of all good design practice. Miss out this chapter and I (or your patients) may come back and haunt you!

So, if you are an established design engineer, an entrepreneur, or a surgeon with a brain wave this book will help you take your idea to the next level. It may simply enable you to communicate with a designer with a better outcome, or it may help you take your product to FDA clearance to market – it may even reduce your time to market. While I would love to claim that use of this book would mean that all of your designs would meet every regulatory requirement, I cannot. What I can say is that it will give you the ability to make sure that you know which ones you have to meet; and it will give you the toolkit and the basic design principles to be able to meet them. Of course I would love to hear of your successes. I am sure emails to the publishers, via their website, will get to me. Please do send comments or suggestions for improvements and case studies, as I would be delighted to include these in the next edition.

To close this preface, I wish to reiterate my aim. This book is targeted at those who wish to design a medical device for sale within the global medical devices community – be that a simple scalpel or an MRI scanner. It is intended to be a reference text that will be on your desk, right next to your iPad and cellphone. Ah, that reminds me, I must apologize to those in the UK. The publishers of this book are USA based, so if you see a "z" where there should be an "s," or if a "u" is missing, or I say cellphone instead of "mobile phone," then I'm sorry but that is the way of the textbook market nowadays. Equally, for those of you in the USA, if you do not follow some of my footnotes then treat them as British idiosyncrasies and laugh out loud (as my editor did!). However, wherever possible I have tried to cross the pond, as they say, by including US English, UK English, and EU cross-references (wherever possible). I hope they work.

Good luck with your designs – may they make patients feel a lot better and make your bank balances a lot healthier.

PJO
Staffordshire, England
June 2012

Acknowledgements

I would like to thank the following for their kind assistance in the production of this book by allowing me to use photographs, case studies and general information: Desoutter Ltd, Intelligent Orthopaedics Ltd; Riverside Medical Packaging Ltd; and Staffordshire University. Furthermore, the U.S. Food and Drug Administration and the Medical and Healthcare products Regulatory Agency (UK) require acknowledgement as their respective online repositories were a mine of information. I also thank all those who have given me advice along the way especially Prof. Peter Thomas, Mrs. Susan Hartman; and (of course) Fiona – my editor, and Renata my Project Manager. If I have forgotten anyone please forgive me.

## CHAPTER 1

Introduction

In most people's vocabularies, design means veneer. It's interior decorating. It's the fabric of the curtains or the sofa. But to me, nothing could be further from the meaning of design. Design is the fundamental soul of a human-made creation that ends up expressing itself in successive outer layers of the product or service.

Steve Jobs (2000)

1.1 What Is Design?

The word "design" causes confusion in every circle of life. One can use "design" as a noun ("This is my design"), as a verb ("I am designing"), and even worse, as a question ("Are you the designer?").

"Design" comes from the Latin *designare* which means "to mark out, point out, describe, contrive." Its form as a noun is the source of a common misconception: "it is *a design*" is normally attributed to a pattern or an image. If you stop an average person in the street and ask them to describe a designer, they are more likely to talk about wallpaper, clothes, hats or tableware than someone who designs, say, a total hip replacement. In this text we are more concerned with the verb "to design": the act of designing, the act of *contriving* and communicating the *contrivance*. The phrase "to design" is very important. It suggests some formality, some structure, and some rigor.

We recognize a contrivance as a *product*: something we are going to sell to someone else. In this sense the product could be an item, a piece of software, or a service. Practically, it is virtually impossible "to design" something that is not intended to be useful to someone. Indeed, being able to design is what makes humans so … so human. We are able to manipulate our surroundings to make them better for us, and we can do some pretty wonderful things with items that are, inherently, rocks and trees. Hence, I propose, that designing is in our blood…it comes as second nature to the human race. But as with other things we humans do, some of us are naturally good at it and some of us are not.

It is a common tendency for us to "hack" – that is, we undertake the activity without any plan or thought. Hence we hack at the problem, just like hacking at an overgrown bush with an axe. Hacking achieves an outcome, but that outcome is always pretty awful and wastes loads of energy. Hacking is also an illegal activity associated with breaking into high-security

Medical Device Design.
DOI: http://dx.doi.org/10.1016/B978-0-12-415822-1.00002-7
1

computer systems; so, to coin a phrase, "neither hack nor a hacker be."[1] The aim of the phrase "to design" is, therefore, a reminder to us all that we need to plan and think about the problem before we start, and definitely not "hack."

So, what are we to deduce from the paragraphs above? Firstly, design is a creative activity – it always ends with something new. On its own, however, this is not design – it could also be a work of art or the breaking of a world record. Hence we need to add something else. That *something else* is a demand, or a need – someone, somewhere wants this *thing*. Again this could still be a work of art or a new world record. We still need to add one other thing: a planned structure, a route map, or a planned process. This now discriminates the activity from all others. If you design an artifact properly, someone somewhere will definitely like it and it *will be sold*. And this is why the word "design" is often confused with the arts – it *is a creative process*. But creativity without structure is not design. As designers we need to harness our creative juices, stimulate every analytical fiber in our bodies, and use every one of our senses to detect opportunities. But we need to do this within an overall structure to ensure that the final outcome meets "a need."

It is clear that we now need to define *the need* as this seems to be core to any design activity – and so it is. Design is about producing something that someone wants. They may not know they want it, but they do want it. How many times have you been to a clothes shop looking for a particular tie, blouse, or shirt? You will have a picture in your mind of what you want – that is your *need*. If the clothing designers have been clever they will have anticipated your need and you buy it. Normally, however, we walk from shop to shop and go home unhappy with something close, but not quite right. So many consumer goods are based on this "predictive" design process – the designers forecast what the consumer will need in, say, a year's time (based on market research) and devise what the need *will be*, or even create a need through fashion trends (how many fashion shows and magazines are there?). Another example of this type of *prospective need* is when technology is rapidly developing and a brand new concept comes from nowhere, e.g. the Sony Walkman. This is often called *disruptive technology* (a new technology that completely changes the way things are done). The Walkman was a success because there was a brilliant forecast that the consumer would want this – but how many disruptive technologies wither on the vine? We do not hear about them *because* they failed – Sinclair C5 anyone?[2]

There is, of course, the immediate response to a demand from the consumer. This is *synchronous* or *immediate need*. This is where the designer is, physically, asked to develop a design based on a direct request. Classic examples of this are buildings – if you ask for your

---

[1] Adapted from Shakespeare, *Hamlet*, Act 1, Scene 3.

[2] The Sinclair C5 was a personal electric vehicle, too far ahead of its time to be accepted then but which now would be applauded for its "green credentials" (Wikipedia, 2011).

own house to be designed you will be working so close with the designer/architect that the information flow is, effectively, synchronous. This is the hardest to manage as the results are often needed immediately too.

The reverse process, *retrospective* or *evolutionary need*, is another name for evolutionary design. It is based on previous designs that may need improvement by implementing a small change. It is nearly always based on an existing concept but a small change makes it different, more desirable, or counteracts an issue. It is often based on customer feedback.

The final form is *scavenging need*. This is like watching vultures or carrion crows feeding on a dead animal. Here the need is *not* to produce something new but to produce something similar. This is often seen in the fashion industry and in consumer goods by those who follow but do not lead. This is commonly known as *me too*. It is said that imitation is the best form of flattery – one would rather be "the flattered" then "the flatterer." It is, of course, intrinsically linked with copying and counterfeit.

Defining the need is, clearly, very important. We will be examining *the need* and how to articulate it in more detail in subsequent chapters. This may not seem to be a creative process, but it is. More often than not, this is the hardest part of the design process. It is hard because we need to understand the customer, fully. But, and this I can promise, if done properly the rest of the process will be so much easier.

So – is defining the need *design*? The previous paragraph should have pointed you to the fact that it is the start of the process. What follows next is the highly creative phase called *ideas generation*. We then have to select a *winner* from the plethora of ideas we have generated; and then we need the *detail (or the embodiment)*. Only when that is done do we have the makings of a design. But, as we will see later, we have not finished, as we still need all of the other elements: packaging, instructions for use, etc., etc., etc. Only when all of this is complete do we have *a design*.

So, *what is design*? The simple answer is:

> *It is a process that takes a need and produces the solution that fulfills said need.*

And, what is a design?

> *It is the solution.*

Unfortunately we now come to the hard bit. We need to put all of the rhetoric into action; and as with most human activities it is easier said than done. The rest of this book is here to make this a lot easier for you.

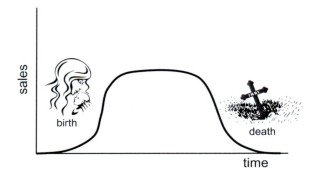

Figure 1.1
The classic life cycle bathtub.

1.2 The Design Life Cycle

People like to talk about the *life* of a design. To have a life means something has to be brought to life and then die; clearly this is a little too anthropomorphic for comfort. But the idea of a cycle is very important and the product's "conception," "birth," and ultimate "end of life" are very pertinent to the designer. In the past the cycle was only concerned with the time to bring new versions to market (revision cycle). Nowadays we consider the whole cycle from manufacture to disposal for obvious environmental reasons. Hence we have two cycles to consider: *revision cycle analysis* and *life cycle analysis*. To avoid confusion between the old and the new, we will use these two terms from now on.

Revision cycle analysis is concerned with keeping products and services "up to date," and this is usually reflected by sales figures. A classic *bathtub* curve describes this cycle (Figure 1.1). At the beginning a new design generates new interest and the sales grow. Eventually these sales plateau as market penetration is reached: consumers get bored, new competitors come along, or there is no one left to sell to. Hence this design's life has ended. To carry on with this version would be silly to say the least. Hence designers need to plan revisions into the process to maintain the sales plateau (Figure 1.2). It is important to understand why a design is never finished … it goes on and on, continually improving, continually getting better. In *quality management* this is called a *continual improvement process*. It is clear that unless the designer is embedded into the discipline into which their design will reside all will fail. We must keep in touch with the *end-user*. We shall see, later, that *postmarket surveillance* is an essential part of this process.

Life cycle analysis has grown in popularity since we all became aware of our environment. The days have gone when any designer can just ignore waste and the consequences of waste. Many products now must have carbon footprint assessments. But as designers we must all be aware of our effects on the environment; the medical devices industry is not immune from this requirement. We will see later that bringing together a design team enables waste to be minimized.

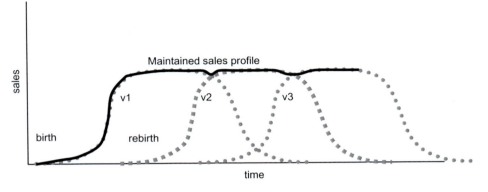

Figure 1.2
Sequential "bathtubs" maintaining the sales profile.

1.3 Medical Devices Definitions

As discussed earlier it is important to be embedded in the discipline in which you practice design. Hence car designers are embedded within the automotive industry: they probably play with their own car at night and on weekends; almost certainly watch or take part in motorsport; and will read every car magazine under the sun. So it is with medical device designers: we must tinker, read, observe and play … but it is unlikely that we will ever be able to use our designs. Hence we operate (excuse the pun) remotely, but we also know that one day our design may just come and save us too. The lesson here is that we need to know as much about the *end use* as the end-user themselves. In fact we need to know more than the *individual* end-user; we need to consider *all* end-users. I can promise you that if you understand your discipline well then your designs will be good and you will get great satisfaction in knowing that you have saved someone's leg, eye, or life.

However, the corollary is that you can also contribute to the loss of someone's leg, eye, or life. To this end medical devices is one of the most highly regulated arenas to work in. Because of this the first staging post is to fully understand what a *medical device* is. Believe me when I say that this is one of the hardest battles you will have with your end-users. Each one will think they are a special case and are, somehow, excluded – they are not and neither are you. More importantly, you are the one who will end up in court.

The European Union (EU) and FDA have tidy definitions of a medical device. Within Europe this is laid down in law under the Medical Devices Directive 93/42/EEC as amended (most recently) by 2007/47/EC. It is by no means concise, but it is neat.

> *(a) "medical device" means any instrument, apparatus, appliance, material or other article, whether used alone or in combination, including the software necessary for its proper application intended by the manufacturer to be used for human beings for the purpose of:*

– diagnosis, prevention, monitoring, treatment or alleviation of disease,

– diagnosis, monitoring, treatment, alleviation of or compensation for an injury or handicap,

– investigation, replacement or modification of the anatomy or of a physiological process,

– control of conception,

and which does not achieve its principal intended action in or on the human body by pharmacological, immunological or metabolic means, but which may be assisted in its function by such means; (EC, 1993)

I would propose that this is pretty clear; if the device is to be used in any clinical environment, on humans, then it is a medical device. Note it does not say that the device has to be in a hospital or used by a clinician – it is defined by *intended use*. Compare this with the equivalent definition from the USA (taken from the Federal Food, Drug, and Cosmetic Act – FD&C Act):

The term "device" (except when used in paragraph (n) of this section and in sections 301(i), 403(f), 502(c), and 602(c)) means an instrument, apparatus, implement, machine, contrivance, implant, in vitro reagent, or other similar or related article, including any component, part, or accessory, which is—

(1) recognized in the official National Formulary, or the United States Pharmacopeia, or any supplement to them,

(2) intended for use in the diagnosis of disease or other conditions, or in the cure, mitigation, treatment, or prevention of disease, in man or other animals, or

(3) intended to affect the structure or any function of the body of man or other animals, and which does not achieve its primary intended purposes through chemical action within or on the body of man or other animals and which is not dependent upon being metabolized for the achievement of its primary intended purposes. (FDA, 2004)

The only difference between the two is the inclusion of *use on any animal* in the USA definition. Note that both clearly distinguish between a device and a pharmacological agent.

Often devices need accessories or additional items for particular functions; these are covered too:

(b) "accessory" means an article which whilst not being a device is intended specifically by its manufacturer to be used together with a device to enable it to be used in accordance with the use of the device intended by the manufacturer of the device; (EC, 1993)

Or if your design is to be used *with something* from definition (a) then it too is a medical device. Again, this is pretty clear. What about things used in a laboratory for assessment of things taken from the human body, and not necessarily in contact with said body? Once again it's covered:

> *(c) "device used for in vitro diagnosis" means any device which is a reagent, reagent product, kit, instrument, equipment or system, whether used alone or in combination, intended by the manufacturer to be used in vitro for the examination of samples derived from the human body with a view to providing information on the physiological state, state of health or disease, or congenital abnormality thereof; (EC, 1993)*

No escape there either – this is a medical device too. Once again the wording is pretty clear – and not worth arguing over. The perennial question comes from the large "made for the patient" market. Some see this as an escape clause or a loophole; it is not.

> *(d) "custom-made device" means any device specifically made in accordance with a duly qualified medical practitioner's written prescription which gives, under his responsibility, specific design characteristics and is intended for the sole use of a particular patient; (EC, 1993)*

What it does do is allow for custom-made devices to *exist* without the premarket assessments required for mainstream devices. Many custom devices could not exist without this definition; it does not mean that the design rigor is any less stringent. It definitely does not mean the medical practitioner takes the blame for any problems – they are not designers or engineers; you still carry the can for any design issues.

The next definitions relate to the power source, or higher risk functions:

> *"active medical device" means any medical device relying for its functioning on a source of electrical energy or any source of power other than that directly generated by the human body or gravity; (EC, 1990)*

> *"active implantable medical device" means any active medical device which is intended to be totally or partially introduced, surgically or medically, into the human body or by medical intervention into a natural orifice, and which is intended to remain after the procedure; (EC, 1990)*

Once again these are very clear. I hope you noticed that it is the distilling of the legal wording into small chunks that has made these definitions understandable. Seeing them on a single sheet is daunting. Throughout this book I aim to present the important issues in this manner.

The lesson, however, is that no matter which of the above definitions your device falls into then do the right things and follow a structured procedure. It is important to note that they are all *medical devices* and are all governed, ultimately, by that definition.

However, as technology advances some products and processes become more dangerous than they were before. So there are two issues we designers have to consider: the first is that we need to keep in constant touch with these important definitions; and the second is that we need to keep up to date with advances in medicine and technology.

There are a number of other requirements you need to be aware of too. In medical devices there are subdisciplines that you need to make sure you work with (EC, 1990 and 1998). The first area is Active Implantable Devices (90/385/EC); within Europe this has its own directive. The second area is In-Vitro Diagnostics (98/79/EC); again, in Europe this has its own directive too. Your product may also fall foul of other directives, such as Electromagnetic Compatibility (EMC) or even Powered Tools. Hence the rule here is have a copy of any legislative document you may think you will need in hand; they are free to download from the websites so there is no excuse. A list of important links can be found in Appendix A; however you should make sure you keep your own list and keep it up to date as the documents change, rapidly. Then, you can omit or include them in the design process, at *your* leisure. We will be looking at this in more detail when we cover *classifications* – that is, putting your device within the correct pigeonhole for regulatory requirements. In the end, as long as you do your design work correctly the classification is immaterial, as a designer should treat every design with the same care. However, it does influence some of your decisions.

It is important now to introduce you to the *police forces* of medical devices. It is worth thinking of them as police because they have the power of life and death over your device and your company.

In Europe each country has its own government body called the *Competent Authority*, as presented before. Even though they are separate bodies they act as one so that an application for a CE mark (the license to sell with the European Union) within one country counts in all of the other member states. The process is somewhat confused by the next level called the *Notified Body*; these are legal entities who are licensed by the Competent Authority to do the CE marking process. These are the people that an applicant would speak to and would be audited by. This is completely different in the USA; here the body is the U.S. Food and Drugs Administration (FDA) and the relevant subset is the Center for Devices and Radiological Health (CDRH). The applicant talks directly to the FDA (via CDRH) and obtains a *clearance to market* (DO NOT use any other words).

It is important to know that the applicant/holder of the CE mark or FDA clearance to market is universally known as the *manufacturer*. They are the top of the regulatory food chain and are ultimately responsible for the *safety* of said device; the designer, the subcontractor, the packager, etc. are subservient to them. However, all are a part of the regulatory process (as we will see later). With manufacturer status come levels of responsibility beyond just insurance such as *reporting, vigilance, postmarket surveillance, gathering clinical evidence*, and much more. Some levels are beyond the scope of this book, but they will be referred to as necessary.

Canada has its own level of complexity (with CAMDCAS), and Japan's is even more complex. However, one has to accept that the application processes are as different as they could possibly be – but at the end of the day the application is about how *you* present your design to the authorities; how you do *your* design is always the same, wherever you are!

At this stage it is worth discussing liability. Ultimately it is the manufacturer who has the ultimate responsibility for liability. However, as with any other discipline, their insurers will try and pass liability down the food chain. Hence it is important, when acting as a subcontractor, that you have relevant insurance in place – and that you never exceed your own level of experience. But, as the designer, one is the hub of the activity. Without the designer nothing happens, no device exists, and there is nothing to present for sale. Hence the product lives and dies at your hands, so your knowledge of the medical devices regime is fundamental. Your adherence to the structured design processes is essential, and your communication with others is of paramount importance. That is why Figure 1.1 includes the image of a newborn baby; this is just how the medical device designer must picture their device. It must be treated with the care and diligence one would apply to a newborn baby; it is, after all, *your* baby.

1.4 Summary

In this chapter we saw how design and continual improvement are essential to medical device design. We were also introduced to the main policing bodies and regulations that control the medical device environment. By now you should be aware that this is a highly regulated arena, and as such any design work has to be up to the mark and should not be approached lightly, frivolously, or with gay abandon. Therefore you now have a few tasks to complete this chapter:

Task 1: Locate and familiarize yourself with the FDA website and your country's EC *competent authority* site.
Task 2: Download the Medical Devices Directive (with amendments).
Task 3: Download the In-Vitro Diagnostics Directive (with amendments).
Task 4: Download the relevant FDA Food and Drugs Act sections (21 CFR 800 series)
Task 5: Read them!

References

European Community (1990). *Active Implantable Devices Directive*. 90/385/EC.
European Community (1993). *Medical Devices Directive*. 93/42/EC.
European Community (1998). *In-Vitro Diagnostics Directive*. 98/79/EC.
FDA (2004). *Federal Food, Drug, and Cosmetic Act (FD&C Act) CFR 21*.
Jobs, S. (2000). Apple's one dollar a year man. *Fortune*, 24th January.
Wikipedia (2011). *Sinclair C5*. <http://en.wikipedia.org/wiki/Sinclair_C5> cited 16.05.11.

## CHAPTER 2

Classifying Medical Devices

2.1 Introduction: Why Classify?

There are a number of reasons why classification is important. The first concerns the patient. Clearly the more risk a given device poses to the patient the greater the reassurance that needs to be given. Not all devices pose the same risk; consider for example a pacemaker compared with a support bandage. It is quite clear that the pacemaker poses the greater risk and, thus, should have more stringent controls attached to its design, manufacture, and eventual sales. The second reason concerns the manufacturer; why should the registration process be as stringent for the support bandage as compared with the pacemaker? The third reason concerns the regulatory bodies; the classification indicates to them the level of risk to the patient and the nature of the beast they are dealing with, and hence the amount of effort they need to put into the *control*. After all, governments have limited budgets too and they need to target their resources at the devices that pose the greater risk.

It is obvious that the prime importance is *safety*. Although patient safety is a priority, one must not forget the users; their safety is just as important. All devices have to obey the prime criteria *do no harm*. But it is also clear that, for some devices, to achieve a clinical goal some harm is inevitable. The question posed is, "Is the risk acceptable?" Consider, for example, a hypodermic needle. Which is the greater harm, producing a small puncture wound in the skin, or not receiving a vaccination? Risk and doing no harm is a balancing act. Classification allows all participants in the regulatory process to understand the risk the device poses. In general, the higher the classification the more chance that the device *could* do some harm. Hence it should come as no surprise that things such as replacement heart valves are high classification and inserts for shoes are low.

Table 2.1 attempts to illustrate how risk and classification are interlinked. Both the USA and EU classifications are illustrated (do not use this for cross-referencing, it is only indicative). The table demonstrates that low risk devices are Class I, and high risk Class III (note in Canada there is also Class IV).

2.2 Classification Rules

Each regulatory authority has its own set of classification rules. In the USA these are stipulated within 21 CFR Part 860 – Medical Device Classification Procedures (FDA, 2010).

Medical Device Design.
DOI: http://dx.doi.org/10.1016/B978-0-12-415822-1.00002-7
11

Table 2.1: Classifications in the USA and Europe

Risk	Low Risk			High Risk
In Europe the classifications are:				
EU	I	IIa	IIb	III
In the USA the classifications are:				
USA	I	II		III

Within the EU they are stipulated within 93/42/EC Annex IX (EC, 1993). Both documents are freely available on the World Wide Web, as are those for any other country. As a medical device designer you must have an up-to-date copy of the classification rules to hand at all times. There is a fundamental difference between the two systems. In the USA classification is by *precedent* and is undertaken by the FDA; that is, you have a classification by comparison with decisions already made by a panel. In Europe there is a long list of questions to answer and *you* decide on your classification. However, if you try and cheat the system by "under-classifying" you will fail.

The classifications are based on *risk to the patient* (as illustrated by Table 2.1). As we will see later, risk is a very important aspect of design control; understanding risk to the patient (and to the clinical team/operator) is fundamental. The FDA process is not a good vehicle to help us understand patient risk and classification. We will use the EU model to understand how classifications are made. For this exercise you will need a copy of 93/42/EC Annex IX at hand.

Table 2.2 illustrates how the EC medical devices rules define the classification of a device. The symbols indicate that this particular rule defines this particular class of device. Hence, if you think your device is likely to be Class III it will fall into the definitions described by one of Rules 6, 7, 8, 13, 14 and 17. Do not assume that these are the only rules you need to consider; all rules need to be examined to make sure that our classification is correct.

Before we go further we need to understand some definitions:

Invasive device: a device that penetrates inside the body either through an orifice or through the surface of the body.

Surgically invasive device: any device that enters the body other than through an established body orifice.

Transient duration: continuous use of less than 60 minutes.

Short-term duration: continuous use of not more than 30 days.

Long-term duration: continuous use of greater than 30 days. (EC, 1993)

Table 2.2: EC Classification Rules in Comparison with the Classes They Define

Rule / Class in rule	Class I	Class IIa	Class IIb	Class III
# 1	☒			
# 2	☒	☒		
# 3		☒	☒	
# 4	☒	☒	☒	
# 5	☒	☒		
# 6	☒	☒	☒	☒
# 7		☒	☒	☒
# 8		☒	☒	☒
# 9			☒	
# 10		☒	☒	
# 11		☒	☒	
# 12	☒			
# 13				☒
# 14			☒	☒
# 15		☒	☒	
# 16		☒		
# 17				☒
# 18			☒	
Hip, knee, and shoulder replacements				☒

Figure 2.1
A typical 3.2 mm bone drill. *(Courtesy Intelligent Orthopaedics)*

These definitions are almost universal, so you should have these emblazoned in your memory. All regulatory requirements come with definitions; the documents are legal documents so definitions are mandatory. If you have ever seen a legal document you will note that the first few pages are *definitions*. They are required so that everyone, as it is said in business school, *"sings from the same song sheet."* The USA and Europe each have their own particular wording so it is important that you keep abreast of them.

2.3 Classification Case Study

To understand classification and risk we will use a case study. For the purposes of this case study we will use the humble orthopedic drill bit and an orthotic shoe insert as our examples. Before we go any further we need to understand the devices. Figure 2.1 illustrates a typical 3.2 mm drill bit. It is used for drilling holes in bones. The process takes just a few minutes and the devices can be reused until they are blunt. They are supplied nonsterile.

Figure 2.2
A typical orthotic shoe insert.

Figure 2.2 illustrates a typical orthotic insert. These are foam-based structures that are inserted into, say, the heel of a shoe to correct gait (walking pattern). They are mass produced and are supplied nonsterile.

2.3.1 EU Classification

The EU process is a step-by-step, rule-based process. At this stage we will start at Rule #1 and work our way upward. Later we will see an alternative method.

> *Rule 1: All non-invasive devices are in Class I unless one of the rules set out hereafter applies.*

Basically, the rule asks us the question, "Is it invasive?" Clearly, being invasive is risky, so one would expect invasive devices to be high risk. A noninvasive device must be less risky… or it should be. The first part of this rule essentially says "all noninvasive devices can be assumed to be non-risky" – the second part says in essence "but if we have evidence that its particular use is risky then we have the right to make the classification higher." Do not stop here! It is tempting to, but we must work our way through all of the rules.

> *Q: Is our drill invasive or noninvasive?*

> *A: The drill enters the body through a surgically made incision (there is no other way to get at a bone). Hence it is invasive. It is NOT a Class I device.*

Hopefully you get the idea. Classification is very simple if you follow the steps. It is a step-by-step approach. In some cases, where you have not fully developed your ideas, doing a classification analysis actually *helps* you to develop a specification. Thus it is worth doing at the start. The classification may change later but we can address that later too.

Let us now consider the multipurpose orthotic shoe implant.

> *Q: Is it invasive?*

> *A: An emphatic no. Hence Rule 1 states it is Class I.*

But we must wait till later to check if other rules cause the classification to change.

I will not go through each rule one by one as this would make the text really boring. But *you* need to read the rules completely and follow them through. It is important to note that it is the rule that classifies the device that we need to concern ourselves with. The rules that do not apply are, ultimately, of little concern. However, we will see later that monitoring this process and recording our judgment is very important.

Rules 2, 3, and 4 are concerned with how noninvasive devices are used. Rule 2 is about storing blood, etc.; rule 3 is about modifying body fluids; and rule 4 concerns contact with injured skin (wound treatment for example). I think you will all appreciate the risk of storing blood for transfusion! The bone drill is invasive so these rules do not apply (are you getting the idea?). The orthotic insert is noninvasive but its use does not fall into any of the three rules so it is *still* Class I.

Rules 5 and 6 are concerned with how invasive devices are inserted. Rule 5 concerns entry via normal body orifices (e.g., a proctoscope); rule 6 concerns entry via a surgically produced entry window (e.g., an arthroscope). Clearly the orthotic shoe insert does not enter an orifice; but the drill does. Rule 6 (modified by 2007/47/EC) states:

> *All surgically invasive devices intended for transient use are in Class IIa unless they are:*
>
> *– intended specifically to control, diagnose, monitor or correct a defect of the heart or of the central circulatory system through direct contact with these parts of the body, in which case they are in Class III,*
>
> *– reusable surgical instruments, in which case they are in Class I,*
>
> *– intended specifically for use in direct contact with the central nervous system, in which case they are in Class III,*
>
> *– intended to supply energy in the form of ionising radiation in which case they are in Class IIb,*
>
> *– intended to have a biological effect or to be wholly or mainly absorbed in which case they are in Class IIb,*
>
> *– intended to administer medicines by means of a delivery system, if this is done in a manner that is potentially hazardous taking account of the mode of application, in which case they are in Class IIb*

Is the drill surgically invasive? Yes. Is it transient use (<60 min)? Yes. Does it fall into any of the other five subrules? Yes. Could it be classed as a reusable surgical instrument? Well logic suggests that drills are sharp when used the first time and get blunt after a few uses, but they can be washed, resterilized and are reusable until blunt, broken, or bent. Our drill bit is reusable so it falls into *reusable surgical instruments*; it is Class I. We check with the agreed definition of a reusable surgical instrument:

> *An instrument intended for surgical use by cutting, drilling, sawing, scratching, scraping, retracting, clipping or similar procedures, without connection to any active medical device and which can be reused after appropriate procedures have been carried out. (EC, 1993)*

Clearly the drill bit fits this definition and it *is* a Class I device.

If the drill was intended to be single use only it falls into Class IIa, but you will need to justify why it is single use as this incurs extra cost to the healthcare provider. Note how the phrase *intended use* is very important.

Rule 7 takes into account longer durations of use. If the use of the drill was for longer than 60 min then its use would change from transient duration to short-term duration. Hence the potential for risk to the patient increases and as such the class increases, but not enough to change the standard classification. But note that the subrules have changed as their potential risk increases with use.

Rule 8 takes this one stage further and puts implants and long-term surgically invasive devices into Class IIb because their risk is greater. Note that this has been a matter of much debate and has resulted in reclassification. The directive 2005/50/EC reclassified hips, knee, and shoulder replacements as follows:

> *By way of derogation from the rules set out in Annex IX to Directive 93/42/EEC, hip, knee and shoulder replacements shall be reclassified as medical devices falling within Class III. (EC, 2005)*

This higher classification for these types of device is universally accepted. The term *implant* has been universally adopted for a device that has been designed to remain within the body with long-term duration.

There are a total of 18 rules (19 if you include the derogation above). I do not intend to go any further as the point has been made. There is only one last technicality to include and that is the use of animal by-products. Rule 17 states:

> *All devices manufactured utilizing animal tissues or derivatives rendered non-viable are Class III except where such devices are intended to come into contact with intact skin only.*

The rise of the prion and the fear of transmission of human variant CJD has made this clause ever more powerful (we shall see this later when we look at application processes). As designers we must be aware of our material and process choices. Do not forget that animal tissue is used as the lubricant of injection molding machines and for thread cutting, so while you may not have specified a non-animal-based material, your design and selection of manufacturing process may put your product into Class III without you realizing it. It is, therefore, wise for you to make sure you are aware of all 18 rules.

Thus the classification we make for the drill bit is:

Classification: Class I as defined by 93/42/EC (modified by 2007/47/EC)

Annex IX Rule 6

For the orthotic shoe insert the classification would be:

Classification: Class I as defined by 93/42/EC (modified by 2007/47/EC)

Annex IX Rule 1

2.3.2 USA Classification

Now let us examine how this classification would be undertaken using the FDA process. The idea is to find a *precedent*, or something like your device that has been reviewed before. Let us take each item in turn and demonstrate. The first port of call is the database section of the FDA/CDRH website (http://www.fda.gov/MedicalDevices/default.htm). The FDA hosts a classification database where you can search for your precedents; this is the *product classification* database (Figure 2.3).

All one is required to do is enter a relevant search term. Using the bone drill as an example let us consider the correct terminology to enter. If we use *bone drill* then we will get hand drills

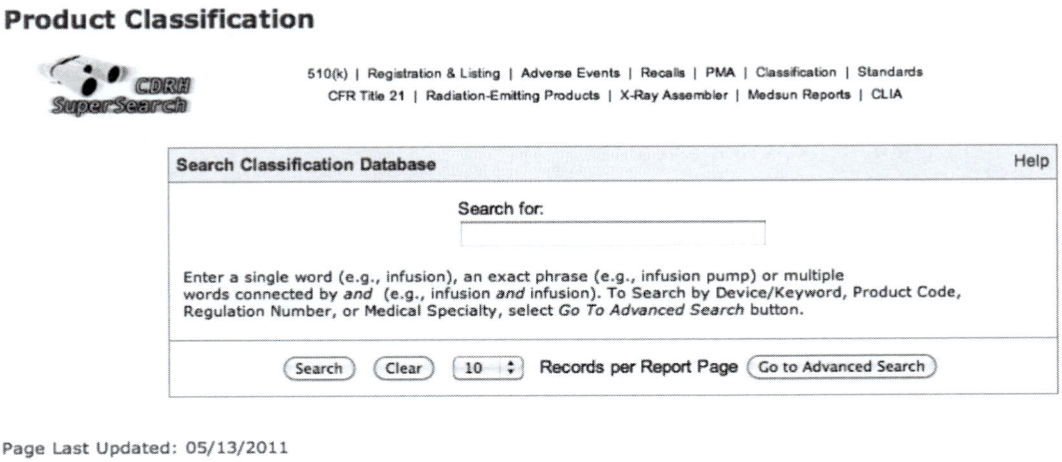

Product Classification

CDRH SuperSearch

510(k) | Registration & Listing | Adverse Events | Recalls | PMA | Classification | Standards
CFR Title 21 | Radiation-Emitting Products | X-Ray Assembler | Medsun Reports | CLIA

Search Classification Database   Help

Search for:

Enter a single word (e.g., infusion), an exact phrase (e.g., infusion pump) or multiple words connected by *and* (e.g., infusion *and* infusion). To Search by Device/Keyword, Product Code, Regulation Number, or Medical Specialty, select *Go To Advanced Search* button.

(Search)   (Clear)   [10 ⬍]   Records per Report Page   (Go to Advanced Search)

Page Last Updated: 05/13/2011

Figure 2.3
FDA product classification database window. *(Courtesy FDA)*

Product Classification

CDRH
SuperSearch

510(k) | Registration & Listing | Adverse Events | Recalls | PMA | Classification | Standards
CFR Title 21 | Radiation-Emitting Products | X-Ray Assembler | Medsun Reports | CLIA

5 records meeting your search criteria returned - *drill bit*

New Search			Export to Excel	Help
⬆ Device ⬇	⬆ Product Code ⬇	⬆ Device Class ⬇	Regulation Number
Bit, Drill	HTW	1	888.4540
Bur, Dental	EJL	1	872.3240
Bur, Ear, Nose And Throat	EQJ	1	874.4140
Bur, Surgical, General & Plastic Surgery	GFF	1	878.4820
Burr, Orthopedic	HTT	1	888.4540

Figure 2.4
Search solution for *drill bit. (Courtesy FDA)*

and power tools; the correct term is a *drill bit*. The results of conducting a search on *drill bit* are shown in Figure 2.4.

A selection of records should appear; all we need do is select the correct one (*Bit, Drill* as highlighted). Following the link takes us to the device classification shown in Figure 2.5.

Figure 2.5 illustrates the official statement. It clearly demonstrates that in the USA the drill bit is Class I. Note that the important bits of data are the classification (I), the product code (HTW), that it is 510(k) exempt, and the regulation number (888.4540). These will become more important once we start to apply to the FDA for *clearance to market.*

The classification for the drill bit would be:

Device: Bit, Drill

Class I. Product Code: HTW. Regulation No.: 888.4540

At the bottom of the table is a further useful piece of information: *recognized consensus standards.* Basically this section is saying that if you use these standards as a part of your design process then you are on the right track – a valuable starting point for any design engineer!

Seeing the FDA process does make one wonder why Europe makes it so hard. It is when a brand new device comes along that the European system shows its true colors. It is down to the manufacturer to do the classification and this will be confirmed once CE marking is

New Search   Back To Search Results

Device	Bit, Drill
Regulation Description	Orthopedic manual surgical instrument.
Regulation Medical Specialty	Orthopedic
Review Panel	Orthopedic
Product Code	HTW
Submission Type	510(K) Exempt
Regulation Number	888.4540
Device Class	1
Total Product Life Cycle (TPLC)	TPLC Product Code Report
GMP Exempt?	No

Note: FDA has exempted almost all class I devices (with the exception of <u>reserved devices</u>) from the premarket notification requirement, including those devices that were exempted by final regulation published in the *Federal Registers* of December 7, 1994, and January 16, 1996. it is important to confirm the exempt status and any limitations that apply with <u>21 CFR Parts 862-892</u>. Limitations of device exemptions are covered under 21 CFR XXX.9, where XXX refers to Parts 862-892.

if a manufacturer's device falls into a generic category of exempted class I devices as defined in <u>21 CFR Parts 862-892</u>, a premarket notification application and fda clearance is not required before marketing the device in the u.s. however, these manufacturers are required to register their establishment. please see the <u>registration and listing website</u> for additional information.

Recognized Consensus Standards

- ISO 13402:1995 <u>Surgical and dental hand instruments -- Determination of resistance against autoclaving, corrosion and thermal exposure</u>
- ASTM F1089-02 <u>Standard Test Method for Corrosion of Surgical Instruments</u>
- ISO 7153-1:1991/Amd. 1:1999 <u>Surgical instruments -- Metallic materials -- Part 1: Stainless steel</u>
- ASTM F 565-04 (Reapproved 2009)e1 <u>Standard Practice for Care and Handling of Orthopedic Implants and Instruments</u>

Third Party Review   Not Third Party Eligible

Figure 2.5
FDA classification statement. *(Courtesy FDA)*

applied for. In the USA, a classification panel makes the decision; the manufacturer has to put forward a case for the classification. A subtle difference, but different nonetheless.

Let us check the FDA classification for the orthotic insert. Searching for *orthotic insert* will not find anything; the results of searching for *orthotic* are shown in Figure 2.6.

New Search   Back To Search Results

Device	Shoe, Cast
Regulation Description	Prosthetic and orthotic accessory.
Regulation Medical Specialty	Physical Medicine
Review Panel	Physical Medicine
Product Code	IPG
Submission Type	510(K) Exempt
Regulation Number	890.3025
Device Class ⟶	(1)
Total Product Life Cycle (TPLC)	TPLC Product Code Report
GMP Exempt?	Yes

Note: This device is also exempted from the GMP regulation, except for general requirements concerning records (820.180) and complaint files (820.198), *as long as the device is not labeled or otherwise represented as sterile.*

Note: FDA has exempted almost all class I devices (with the exception of reserved devices) from the premarket notification requirement, including those devices that were exempted by final regulation published in the *Federal Registers* of December 7, 1994, and January 16, 1996. it is important to confirm the exempt status and any limitations that apply with 21 CFR Parts 862-892. Limitations of device exemptions are covered under 21 CFR XXX.9, where XXX refers to Parts 862-892.

if a manufacturer's device falls into a generic category of exempted class I devices as defined in 21 CFR Parts 862-892, a premarket notification application and fda clearance is not required before marketing the device in the u.s. however, these manufacturers are required to register their establishment. please see the registration and listing website for additional information.

Third Party Review   Not Third Party Eligible

Figure 2.6
Orthotic classification. *(Courtesy FDA)*

Once again, note the relevant data: it is Class I and 510(k) exempt. However, the device is *GMP exempt*. The value of this piece of information will be explored in more detail later. The classification would be:

Device: Shoe, Cast

Class I. Product Code: HPG. Regulation No.: 890.3025.

**Table 2.3: Comparison of Classifications in the USA
and Europe**

Device \ Class	USA	Europe
Reusable surgical instrument	I	I
Disposable suture needle	I	IIa
Bone screw	II	IIa (short duration)/ IIb (long duration)
Blood sample container	I	I
Hip prosthesis (implant)	II	III

Do not use this table for classification purposes, it is only for comparison.

Table 2.3 illustrates some comparisons between classifications of devices in Europe and in the USA. By and large (apart from IIa and IIb) the classifications match. But do not let this fool you, you must check the classification and not simply assume.

…when you assume, you make an ASS out of U and ME[1]

2.3.3 Special Cases

There are two special cases to consider. The first is custom-made devices and the second is a device intended for clinical investigation. These are defined as follows:

"custom-made device" means any device specifically made in accordance with a duly qualified medical practitioner's written prescription which gives, under his responsibility, specific design characteristics and is intended for the sole use of a particular patient.

The above mentioned prescription may also be made out by any other person authorized by virtue of his professional qualifications to do so.

Mass-produced devices which need to be adapted to meet the specific requirements of the medical practitioner or any other professional user are not considered to be custom-made devices; (EC, 1993)

Basically, this means the device is a one-off. Clearly one cannot go through the full regulatory process for every single one-off device, but this does not mean that the rules of engagement do not apply. Consider, for example, a simple walking frame specifically designed for a patient or, in comparison, a designed-for-patient replacement hip. Both are one-offs but the risks are, clearly, not the same. Due consideration has still to be taken with the design process – this is not a "get out clause." One should still ascertain the device's classification to assess the diligence one needs to apply.

[1] Attributed to Benny Hill – much maligned British comedian – from his television series of the 1970s

Medical devices for clinical evaluation come under a whole different regime. Surgeons, engineers, and scientists often need to conduct evaluations before obtaining clearance to market. This means that the device they intend to use must be cleared for clinical investigation. We will examine this in more detail in later chapters. Once again, this is not a "get out clause" and the diligence required is probably higher than with a commercial device. It is "experimental" so not all of the pitfalls, side effects, or clinical issues will have been ascertained. Therefore the design process needs to be failsafe in its approach.

2.4 Classification Models

One of the big issues with medical devices companies is the number of times they have to apply for a CE mark or for an FDA clearance to market. However, it is rare for a company to step outside of their comfort zone; and even more rare to review the process that took them to the eventual outcome. It is possible to learn from the activity itself, be it a success or a failure. Hence it makes sense to learn from the past, but even better to plan to learn from the *now*. The regulatory steps that were taken to reach the ultimate outcome need to be recorded and analyzed. It is possible to save the company a lot of time by learning from previous successes and previous failures.

Processes govern the whole of the medical devices industry, yet one of the most important tasks (the determination of classification) is often overlooked. It should be the first thing to be done; even at the "Oooh, what about this for an idea?" stage there should be an initial estimate of classification. Hence the idea that all medical devices companies have a classification process is not such a daft idea. One must remember that the costs of administering medical devices grow disproportionately with each classification level; removing an idea, at an early stage, due to offensively large expenditure profile is just as important as developing a new, money spinning idea.

Figure 2.7 illustrates a typical flow chart developed for determining if a device is Class I (within Europe). While this does not match completely with the USA it is worthwhile, if only to understand your idea better or even to second-guess the classification panel. There is no reason why you shouldn't produce one of these for each of the devices you intend to market. At the very least it tracks your decision-making process and can be signed off as a part of your technical file. Note that all of the rules that do not end with a Class I device have been removed.

This flow chart has turned the classification rules on their head; instead of "What is the classification?" the question has become "Is my device Class I?" This is probably the best bottom-up type of approach to take. After all, we would all save loads of money in relation to auditing, administration, and insurance if all or our products were the lowest class possible. There is no credit, no medal, and no prestige in placing your devices in a class too high for their worth. Do not fall into the trap of "I will play safe and make it a higher classification"; this is not the idea and defeats the whole purpose of this chapter.

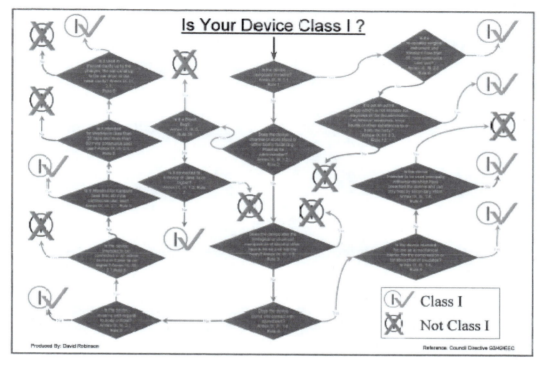

Figure 2.7
Is my device Class I? *(Staffordshire, 2008)*

Table 2.4: Device Classification versus Control Measures

Design Control	Low→High			
EU Class	I	IIa	IIb	III
USA Class	I	II		III
Selfregulation	High←Low			

2.5 Classification and the Design Process

In both the USA and within Europe the degree of control one applies to the whole life cycle of the device increases with classification (as illustrated by Table 2.4).

The FDA use specific terms for the level of control, which are worth remembering wherever you intend to work:

Class I means the class of devices that are subject to only the general controls...

Class II means the class of device that is or eventually will be subject to special controls...

Class III means the class of device for which premarket approval is or will be required...
(FDA, 2010)

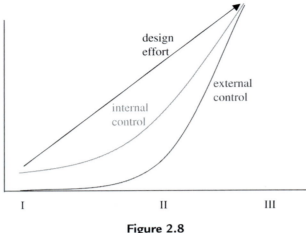

Figure 2.8
Design effort versus classification.

Basically this means that the level of investigation into your design processes before giving any CE mark, or clearance to market, is negligible for Class I products. Indeed it is virtually self-regulated. You may be tempted to think that this means you do not need to do any "real design" but you would be wrong. You have to think about the situation when something goes wrong: for example, when you are placed in court defending your product that has just maimed someone, or when you have to defend your design to an expert designer with no evidence. Self-regulation, therefore, means *you* make sure you have the design files in place.

The higher the classification, the "thicker" your design's file becomes as it will contain more investigations to make sure it is safe to use. The overall process is the same; it's the amount of work that increases. Your company will have to defend its submission to the authorities before it can be placed on the market. The file has to be bulletproof; the higher the classification the bigger and faster the bullets. The people that audit your files are not stupid, they are highly intelligent scientists and engineers so do not try to fool them – you will fail.

Figure 2.8 illustrates the amount of effort required for each classification. The amount of external control is negligible for Class I products, but the company's internal control is significant. For Class III devices the amount of control exerted by external factors is probably equal to (if not more than) that of your company. This does not means your hands are tied and you cannot make any design decisions; it simply means you have to justify them – fully. Hence the amount of effort you must apply to the design process increases too.

External control does not only come from the regulatory bodies. As the classification increases so does the number of *standards* that apply. There are many standard bodies to work with but in general we will be talking about ASTM (American Society for Testing and

Figure 2.9
A common orthopedic power drill. *(Courtesy Desoutter)*

Materials), ISO (International Standards Organization) and BS (British Standards). These are the three standard bodies for the USA, international community, and Britain, respectively. Unfortunately, working with a standard in the USA does not necessarily make this compliant with a standard within the EU; hence the recognized consensus standards become a very good starting point to determine which standards apply.

Also, your device may cross discipline boundaries, one example being a simple orthopedic drill (Figure 2.9). While this is clearly a medical device, because it is powered it also falls under the remit of powered hand tools; it could well be subject to electromagnetic compatibility regulations too, and because it makes a noise it falls under noise emissions regulations. As the designer you must make sure that your device meets all relevant standards and guidelines. Unfortunately Class I designers only find this out when it is too late; Class II, and above, design teams have the failings pointed out to them when they have their first audit. This is the main drawback of self-regulation!

2.6 Summary

In this chapter we were introduced to classifications. We saw that there are two different approaches in the USA and Europe but that they ultimately end in the same place. We saw that the classification in the USA may not directly match that in the EU. We learned how to classify our devices and determined what the classifications mean to our design controls and our development costs. Therefore you now have a few tasks to complete to ensure that you familiarise yourself fully with the classification methodology.

Task 1: Make yourself fully aware of the FDA database of product classifications.
Task 2: Determine the classification (EU and USA) for
   i) A single-use scalpel
   ii) A dental filling (you will need to think to find this)
   iii) An x-ray imaging machine
Task 3: Produce a chart similar to Figure 2.7 for Class IIa devices.

References

European Community (1993). *Medical Devices Directive.* 93/42/EC.

European Community (2005). *Reclassification of hip, knee and shoulder joint replacements in the framework of Council Directive 93/42/EEC concerning medical devices.* 2005/50/EC.

FDA (2010). 21 CFR, Subchapter H, Part 860.

Staffordshire University (2008). *A Systems Approach for Developing Class I Medical Devices.* BEng(Hons) Thesis.

## CHAPTER 3

The Design Process

3.1 Design Process versus Design Control

There are two fundamental reasons why we need to control our design processes. The first is regulatory. In all medical devices regulations (for example FDA, 1997; FDA 21 CFR 820.30; ISO 13485; and ISO 9001) you will find the term *design control* – hence we have to control our design process in order to fulfill our obligations to the medical device authorities. The second reason is for the life of the company. It is very easy to spend (sorry, waste) lots of money undertaking uncontrolled design leading to outputs that are not fit for purpose. This is futile. We should, as medical device designers, work to *right first time every time*. Using the 6σ[1] model, if your first three designs are rubbish, then your next 999,997 have to be spot on. Controlling the process also saves time. This not only saves money (saving staffing costs, etc.) but also leads to shorter *time-to-market*, bringing obvious advantages.

People often wrongly assume that there is a conflict between a process and that which controls it. Any control engineer will tell you this is false. Before you can control anything you need to understand the process – you need to understand how the process changes the *input* to an *output*. You need to measure the input and you need to measure the output. It is the relationship between the two that is the process. Figure 3.1 illustrates a design activity as a typical control block diagram.

The design process illustrated in Figure 3.1 is "open loop": there is no feedback; the output has no influence; and, worst of all, there is no measurement of whether the output is right or wrong. Control engineers correct this by "closing the loop" – by introducing feedback. This is illustrated in Figure 3.2.

Closed loop systems are known to be more efficient (Schrwazenbach & Gill, 1992), but we are not designing a control system for a machine tool – we are trying to examine design as a process. *What is the lesson?*

It is something that the Six Sigma[2] fraternity picked up on very quickly. The term DMAIC (Define, Measure, Analyze, Improve, Control) is a fundamental tenet of 6σ (Bicheno and

---

[1] 6σ: stands for Six Sigma, a popular design/manufacturing management tool.

[2] Six Sigma (6σ) was started by Motorola and spread to worldwide fame with General Electric's adoption: its aim is to minimize defective components to a maximum of 3.4 in 1,000,000. There is a plethora of texts available.

Medical Device Design.
DOI: http://dx.doi.org/10.1016/B978-0-12-415822-1.00002-7

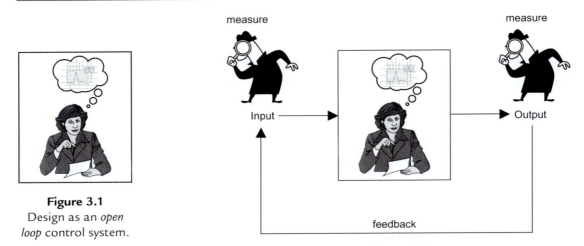

Figure 3.1
Design as an *open loop* control system.

Figure 3.2
Design as a *closed loop* system.

Catherwood, 2005). There is no reason why we should not form the same connection. That is why I introduced the closed loop example to demonstrate the need to measure input *and* output (etc.); the requirements to enable feedback; and the need to understand the process, so that ultimately we can *control* our design *system*.

However, we are very lucky as we are able to define our system; we are able to *define the process*. If we do this correctly we will be able to implement the concept of DMAIC and, ultimately, we *will* control our design process.

It's not all plain sailing. Anyone who has been to a concert and heard the scream from the speakers when a microphone is poorly placed will know the detrimental effects of feedback. To control our system we need to understand it, understand how it responds to changes in data, and understand what "makes it tick," Hence, to understand the design process is fundamental in our task to control it.

It should be apparent that I have proposed and highlighted the need to *control* our design process. In order to *control* it we need to *define* it. Subsequently we must *measure* and *analyze* the results (input and output). In this way we are able to continually *improve*. No wonder Six Sigma picked the letters DMAIC out of the alphabet.

The other thing to remember is that the design process is an *iceberg* (Figure 3.3). Most of an iceberg lies under the water, where it is hidden from sight; so it is with design. Everyone sees the tip of the process. Imagine a brand new, shiny Ferrari – how many people think, when they see that lovely new Ferrari, "Ooh there was lots of work done on the calculations for the strength of the fan belt" – not many! They will all look at the color, the upholstery, the wheels, the engine bay, and the logo. They will not see all of the hard work that lies

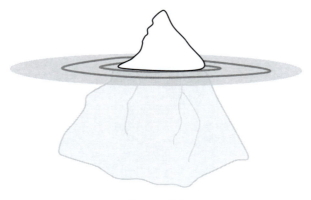

Figure 3.3
The design iceberg.

underneath – all that lies under the water. Eventually your project will be that Ferrari, but you must not forget all that lies under the water. Forget this simple fact and, just like the Titanic, *your* whole design process, and your project, will sink – out of sight.

This chapter, and the subsequent chapters, will present design models and processes that enable you to develop your own design process. It is important for you not to just copy from the text; this will not help you or your company. You should use the following chapters as guides to help you develop your own. Your processes need to fit your company, your discipline, and your aspirations – if they do not then they are as useful to you as a *"chocolate fireguard."*[3]

3.2 Design Models

There are several models of the design process. We shall address them here and put them into the medical devices context. You must remember models change with the wind and, unfortunately, with fashion. So what is an acceptable model today is out of favor tomorrow. However, the fundamental design process is constant. We will examine these models to ascertain the constants and then assimilate these into something we can shape and model to suit our own interests and aspirations.

There are two main models for design in the texts. The first was developed by Pahl and Beitz, the second by Pugh. They are both configured around engineering design. As we are dealing with medical devices this is the most appropriate design philosophy to use. You may ask why. Medical devices are products in their own right, be they software or hardware. At the end of the day they have to be made; they have to be engineered – hence an engineering design philosophy is the most appropriate. Later we shall see that it fits the requirements

---

[3] Old English saying describing anything that is useless – other similar phrases include "like a fish needs a bicycle."

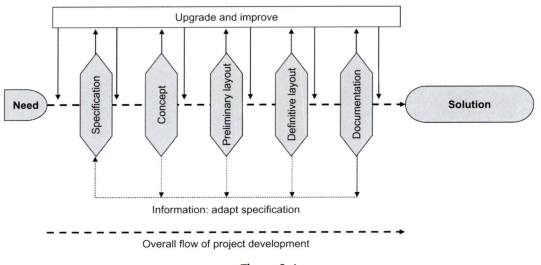

Figure 3.4
Design model adapted from Pahl and Beitz (2007).

of the regulatory bodies (like a glove). However we must still integrate with graphics designers, product designers, etc. and as such we will also look at how to bring these into the "engineering design family."

3.2.1 *Pahl and Beitz, and Pugh*

Figure 3.4 illustrates Pahl and Beitz's[4] model for the design process (originally from the 1980s). It is nearly 30 years old and is starting to show its age. But, the basic concepts are still worthy of investigation. This and the following model by Pugh are linear processes; that is, they fundamentally assume that the process starts at one end and moves in a (roughly) straight line to the final outcome.

Figure 3.4 illustrates an interpretation of the design model proposed by Pahl and Beitz. It is, as described earlier, linear in approach. The very left end is the start of the process and this is universally known as the *need*. There are some who call this the *brief*, but this is a term often used by product designers. In essence they are both the same; they give a *first description* of the concept.

The second phase concerns the development of the first main design document – the specification. The third phase concerns developing a concept and an initial layout, leading to the selection of a single front-runner. This in turn leads to the definitive layouts and finalizing documentation.

---

[4] Pahl and Beitz's text is in continual print – translated from the German original – and is in its third edition.

Where, or when, the need comes from is often a matter of debate…but it always exists. In essence it comes from five main sources (as described in Chapter 1). The first source is a customer who specifically asks for something, or from the marketing department/sales force who will have spoken to their customers – this is *immediate need*; the second source also comes from the marketing department and relates to "copying" someone else – *scavenging need*; the third relates to market research predicting a trend and formulating a *prospective need*; the fourth comes from research and development where a disruptive technology has invented a need for its use – this is also a *prospective need*; the final source is directly from postmarket surveillance and relates to the evolution of the device – *evolutionary need*. The main thing to ensure is that the need is documented and accepted – this documentary evidence of a need may be called a *brief*.

While this seems logical, it implies a "wait and see" philosophy. It suggests there is little space for alterations, changes of mind, and changes of demands. Backward and forward arrows accommodate feedback between phases. The process waits for problems to arise before they are discussed. This can result in eddies: loops in which the designer and their partners get stuck.

Figure 3.5 attempts to illustrate these phases. The first phase can be considered a *clarification* phase. That is, this phase enables the designer (or design team) to make themselves fully aware of the need and the environment in which the need operates. It also gives the designer time to talk to the end-users (et al.). All of this is necessary in order to develop a full specification before going on to the *conceptual design* phase. This phase enables the designer to develop initial ideas from which to select a single design to go through to the *embodiment* phase, where a final prototype is developed. Once accepted, the prototype can go through to design for manufacture (detailed design) and final documentation.

It is now pertinent to introduce Pugh's[5] Total Design model (Figure 3.6). Once again, this is a linear approach. However, Pugh took the concept of a *product specification* to a higher level. He identified that if time is spent developing a good specification then all else will fall into place.

Unlike the original model of Pahl and Beitz, Pugh incorporated manufacturing into the design process. For us this is an important step towards our design model. Nowadays there are terms such as D-4-X and Design-for-Manufacture. These demonstrate how much of the tail has been brought into the design process.

While these models are worthy, in reality they do not give a full picture of actual activity. They visualize the processes but not the activities. To this extent I shall present a didactic model.

---

[5] Pugh proposed the concept of "Total Design" in his seminal book of the 1990s.

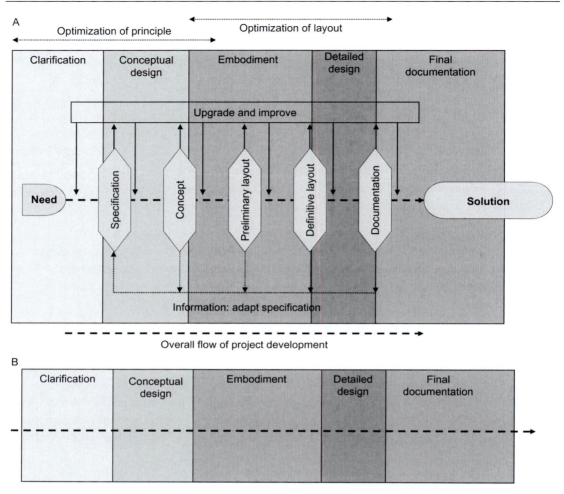

Figure 3.5

Linear design phase model: (A) original model; (B) extracted model.

3.2.2 Divergent–Convergent Model

Because the models are abstract in thought, it is difficult to visualize the reality. I now propose to present an alternative model, which I hope better illustrates the design process in all its glory. If we take the phases identified previously we can produce a model, as illustrated in Figure 3.7.

The overall boundary is "funnel" shaped for a specific reason, which will be described later. Again, this has been presented as a linear task, from left to right, solely for clarity of presentation. Hidden within this "funnel" is a series of twists, turns, roundabouts, and stop signs; again, these events will be discussed and presented in more detail later.

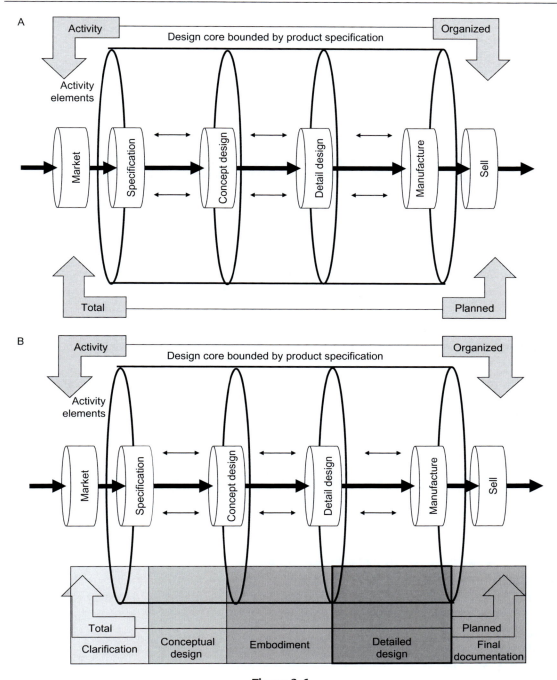

Figure 3.6
Total Design model adapted from Pugh (1990): (A) original model; (B) model adapted to include phases.

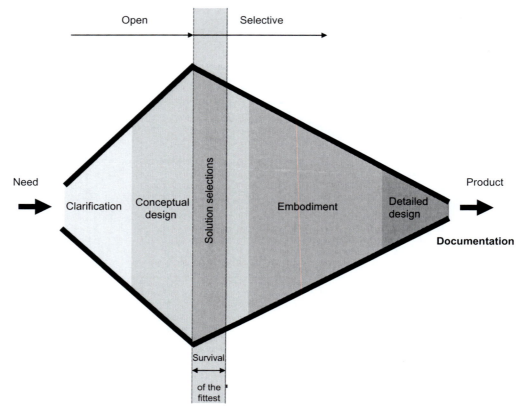

Figure 3.7
Medical devices design model.

Moving from left to right takes us through the stages presented earlier. Firstly a need is established; secondly the full background to the need is clarified; thirdly concepts are developed; and finally detailed design takes us to a final outcome. There are three more phases to consider, which present the overall philosophy. The first overall phase is "open": this means the designer needs to be open to everything – there are no holds barred, nothing is considered to be stupid. This phase only works if the designer is open to suggestion. The middle phase, "survival of the fittest," is a selection phase. Here the designer picks the best option; that is why you need an open phase otherwise you would have nothing to pick from. The third phase is a "selective" phase; here the designer is selective about what they do and the tasks undertaken are often highly prescribed. Figure 3.8 attempts to illustrate this by describing the kind of activity undertaken during the design process. The very start of the process is methodical (producing the *statement of need*), but soon the brain starts to work

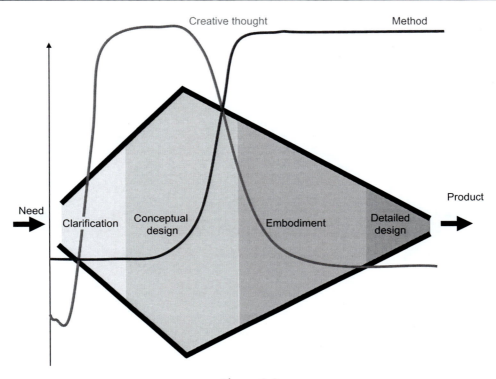

Figure 3.8
Design process versus effort.

and the whole process is dominated by creative thought. You need to be thinking creatively to produce options; you must behave creatively to identify and research options. However, soon the options will dry up and the hard work of selecting a front-runner and then making the front-runner work begins. This is now method dominated and you will find yourself following well-trodden paths.

We now take this model one stage further and look at ideas. Figure 3.9 illustrates this graphically. At the start you will have some initial thoughts about the final outcome. In itself this is dangerous and is called the *sacred cow syndrome*. The sacred cow syndrome is named after the belief in India that cows are sacred and must not be harmed. Often a designer will have an idea that becomes a sacred cow: no matter what happens that idea is predominant and nothing will move it. Some call this *strength of will*, but as designers that sacred cow comes out of the end of the process once we have proved it is the best, not from some belief at the very start.

The number of ideas generated peaks just before the start of the constriction of the funnel; that is because the funnel is supposed to demonstrate how the design process works. At the

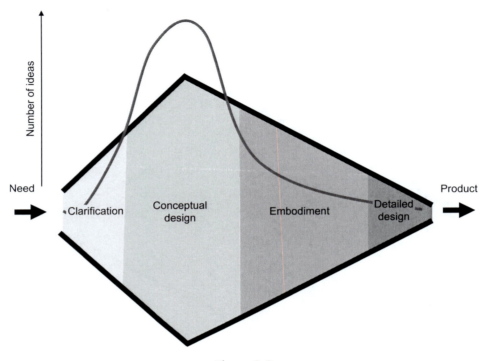

Figure 3.9
Design process versus ideas generated.

beginning a single need produces an abundance of ideas that could satisfy that need – hence the funnel expands to demonstrate the expansion of ideas. Eventually we need to select an idea and develop that into a single solution that fully meets the demands laid down by the statement of need; hence it constricts, demonstrating that we focus on a single outcome. This process is illustrated graphically in Figure 3.10.

At the very left of the process one document is produced: *the statement of need* or *design brief.* This document outlines the demand and gives some indication of the requirements. It is not complete enough to undertake the design process but is detailed enough to make a decision whether to go forward or not. To complete the clarification stage one needs to produce the highly detailed *product specification* or *product design specification (PDS).* To do this you will need to immerse yourself in the discipline, interview end-users, and discuss needs and demands with both your customers and your subcontractors – all are important. As indicated in the diagram, it is likely that the PDS will go through several stages and several draft PDS will be produced before a final version is agreed on. We will cover this in more detail in Chapter 5.

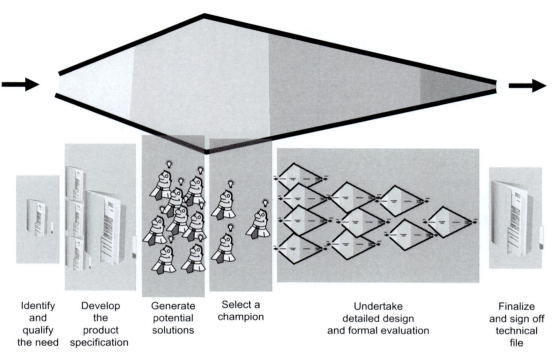

| Identify and qualify the need | Develop the product specification | Generate potential solutions | Select a champion | Undertake detailed design and formal evaluation | Finalize and sign off technical file |

Figure 3.10
Populated design model.

The next stage takes the PDS and expands this into ideas. We start thinking of solutions that meet the requirements of the PDS. As a consequence we should populate the *design space*[6] with numerous ideas, the more the better! Ideas generation techniques are covered later in the book. Figure 3.10 illustrates that from the PDS several ideas emerge. We now need to reduce the space to an ultimate, single champion: the one solution that, above all others, meets the list of requirements stipulated in the PDS. You should be starting to realize how important the product design specification is – it is the one document from which all else leads.

The funnel is converging. All of the expansion of ideas that was allowed and supported in the early stages results in a healthy, robust, and vibrant selection process, one that enables you to select the single idea that stands out amongst all of the others. This is why I prefer to see it as a divergent–convergent model, as it suggests, nay forces, you to be creative, but in a structured and robust way.

---

[6] In the context of this book the design space is defined as the total number of solutions that *all* fit your specification. If you could draw a picture of your specification it would be a multidimensional "surface." Solutions that meet your specification would sit on this surface – other ideas would not and hence are not in your design space.

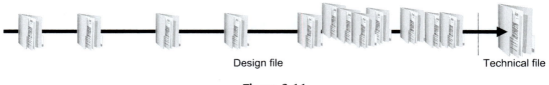

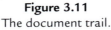

Design file   Technical file

Figure 3.11
The document trail.

In comparison, where the early stages are fun and dynamic, the latter embodiment stage can be seen to be monotonous. This is because we leave the creative element behind once we start to converge. This phase, however, is highly critical as it is the stage where the important decisions are made. In this phase we perform standard tasks and investigations to produce a working design that meets the needs of the PDS. These can be repetitive, but hidden within this monotony is the burningly obvious fact that every single device is actually a composite of a mountain of smaller designs (remember the iceberg), each with their own particular need, each needing a single solution to be selected, and each needing to be designed. Figure 3.10 demonstrates this with numerous little divergent–convergent processes, each indicating their own cog in the giant wheel of your overall project. This takes a lot of planning and project management, and we will be looking at this later too.

All the way through the process you will need to produce a document trail. Figure 3.11 illustrates this and demonstrates how the number of individual documents grows. These reside within the *design file*. This is a "virtual document" as it is probably bigger than any file you have on your shelf. It will probably be a collection of folders, and may even be a whole filing cabinet. The important thing is that the design file records the whole of the design process from start to finish. Every meeting, every decision, and every change must be recorded here. We shall meet this again when we examine the *product approval process*. The whole process culminates with the *technical file*. This is medical-devices-speak for the one document that fully describes your device: how to make it; how it has been developed; how it has been assessed; and how it meets any essential requirements. *Unless you follow a structured design approach you will not be able to produce a technical file of sufficient rigor to get through any medical devices audit.* This cannot be stressed enough – the applicable phrase is, *"Start as you mean to go on."*

3.3 Managing Design[7]

Do not assume that management of design is only a management of process. Design involves people, so it is as much about people management as it is process management. It is beyond the scope of this book to turn you into a complete manager; however some simple tasks are

---

[7]There are whole textbooks just on management of design – it is a subject in its own right. Universities have whole master's degree programs on this topic.

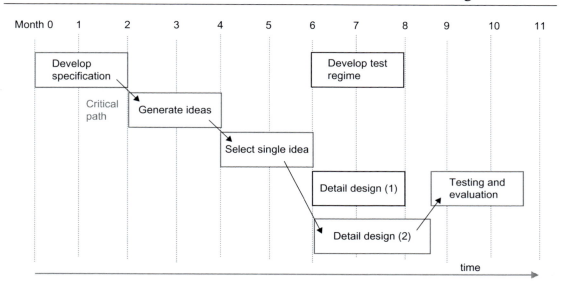

Figure 3.12
Example Gantt chart.

essential. The tasks I intend to introduce, which I see as essential to managing design, are project management, team building, and fully understanding several design management models: serial design, concurrent design, collaborative design, and holistic design.

3.3.1 Common Design Management Models

3.3.1.1 Serial Design

You should have realized that the models presented earlier in this chapter are effectively serial. The tasks happen one after the other and in a laid down sequence (Figure 3.12). Even projects with a multitude of activities have nose-to-tail type lines of communication. The best way to visualize this is the relay race in the Olympics. One team member cannot start to run until the previous one passes over the baton. The baton finally reaches the end, and gets there successfully, but this may not have been the fastest way. For many simple designs, or for very small, micro-companies (1–2 people) this is the only sensible model as only one task can be done at a time. But for larger projects and larger activities it leads to overlong *lead times* (time from start to delivery). There is only one way to speed up serial design and that is strong, robust project management using project management tools such as Gantt charts, PERT, etc.[8] However, although focusing on the critical path will chop some time off the overall project, a serial design model is one that delays progress. Figure 3.12 illustrates a

---

[8]Project management is an essential part of the design process. If you are not au fait with project management techniques stop now and get up to date with this discipline. There are several e-books you can acquire, e.g., http://www.eejp.org/resources/project_management/introduction_to_project_management.pdf

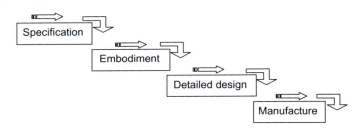

Figure 3.13
The waterfall design model.

typical Gantt chart for a design project. It is by no means a model Gantt chart and should only be considered as an example.

The example Gantt chart illustrates how one task follows another – and this is the essential problem with a serial model. One only identifies an issue once a task has started. In the chart, "testing" take over the baton from design when they have finished. Suppose testing determine something is wrong; this information goes back to design, who change their ideas, and then testing take over the baton again. This backward and forward cycle is common in poorly managed design systems. Also, as has happened in the past, the whole design may have to be scrapped; as you can imagine this wastes a lot of time and money. An example of how project management can speed up the process is by running fully independent tasks simultaneously (develop test regime, detail design (1), and detail design (2)). Even so this is still, essentially, nose to tail. In essence it is hard to "look backward," or to enforce feedback control in serial design. Some people demonstrate this by using a waterfall model (Figure 3.13). The project flows like water from one task to the next by flowing over the edge after said task is complete – and, as everyone knows, water does not flow uphill.

3.3.1.2 Ad Hoc[9] Feedback

To try and overcome the issues discussed in the previous section people have tried to insert feedback into the system. However this feedback tends to be "reactive" and only happens when a problem has been identified.

An example of this form of ad hoc feedback would be when final packaging is selected. Often, at this stage the packagers will say, "If only your device was 12 mm (½ inch) shorter it would have fit in one of our standard boxes." This is clearly true, but unhelpful.

While ad hoc feedback is better than none at all, it is inefficient. It leads to long lead times, often leads to severe frustration, and is a major source of internal tension: the classic example being between the manufacturing office and the design office. It is often the case that the manufacturing office accuse the design office of designing things that cannot be made and the

---

[9] Ad hoc is a Latin word. It generally signifies a solution designed for a specific problem or task. It is not general, and not intended to be able to be adapted to other purposes. It is often a reaction to an event.

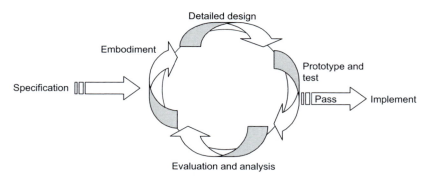

Figure 3.14
Iterative model.

design office accuse the manufacturing office of a lack of creative spark. Neither is true – it is efficient communication which is missing.

Do not think that this is only applicable to large companies. It is a mindset that even the smallest micro-company can get into: the same issues and problems arise.

It should have started to be apparent that communication is an important aspect in the control of design. One of the next steps taken to improve communication is the adoption of concurrent design/concurrent engineering.

3.3.1.3 Concurrent Design/Concurrent Engineering

Concurrent engineering[10] suggests that the waterfall approach and its ad hoc nature are not conducive to an efficient design system. The model recognizes that iterations are required to come to a final solution and that teams will be set up (or selected) to obtain the answers or solutions to a particular problem. It is still serial in nature but, where possible, tasks that can be done at the same time are scheduled *concurrently*.

First, let us examine the change to the waterfall model. Instead of having a drop at the end of the task, there is an eddy that signifies some form of iteration is inevitable and this iterative process involves all of the tasks (Figure 3.14).

Even if you never adopt concurrent engineering as a design model, embrace the concept illustrated in Figure 3.14. Its ultimate goal is to be *right the first time every time*. While each individual task in the loop cannot be right first time, the goal is to use the loop such that the design is not implemented until it is right.

You should also notice that there is an inherent danger in this model: never-ending loops. In computing "getting stuck in a loop" is a common issue – you will have met it on your

---

[10]Once again, this is a topic in its own right. A quick search of e-books will find you a plethora of texts to satisfy your curiosity.

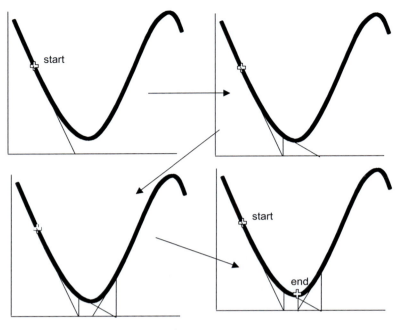

Figure 3.15
Finding a minimum.

personal computer when a program locks for some inexplicable reason; the most common reason is that the program is going around and around a single loop with no exit in sight. This can happen with this model. If not managed correctly, it is perfectly feasible for an idea to be bounced around between departments with a conclusion never being reached. In mathematics this is exemplified by a Newton–Raphson[11] search. Figure 3.15 illustrates a graph of a function; we need to find the lowest point. One way to do this is to pick an arbitrary point and draw a tangent to the graph. Where this crosses the x-axis is the next point, and so on until the minimum is found. Figure 3.15 illustrates this when it works. However Figure 3.16 illustrates the condition when it "gets stuck in a loop."

This type of loop exists in design, particularly between the design team and the actual customer. Quite often the design team will come up with a solution, only to be met with the nightmare scenario "*ah but you've forgotten… .*" The design team then changes the solution to meet this new requirement only to be met with "*well it wasn't that important and it was better last time.*" The solution is changed again and the team members are met with "*well it's true this is good, but on second thoughts the one before was really good.*" This type of loop happens all of the time; good specifications and good management break the cycle.

---

[11] As you may guess this is *the* Newton. Legend has it his first search technique was developed to find books in the university library (before a Dewey system was in place). This is a more advanced approach (Wikipedia, 2011).

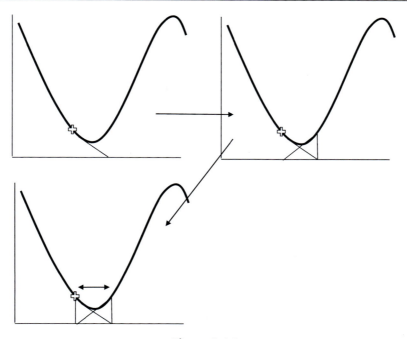

Figure 3.16
Getting stuck in a loop.

Now consider the power drill described in previous chapters. If a company were to design this from the start then the serial model suggests that only one team works on the project from beginning to end. Concurrent design recognizes that this is not the case and that there may be more than one team working on the overall design. Indeed these teams may well be highly specialized. So, for example, one team may be designing the chuck, another may be designing the drive unit, another the battery pack, and a final team the casing. This makes four teams. It is possible for one team to wait for another to finish (as in the serial model), but concurrent design suggests that they can work at the same time. Figure 3.17 illustrates this. Using the relay race analogy presented earlier, it is like giving all four runners a baton and setting them all off at once – the race clearly finishes in a quarter of the time. While in the Olympics this would be frowned upon, in design management it is a blatantly obvious model to use.

For the model presented in Figure 3.17 to work, there needs to be a strong overall project management role. Also, each task needs to be responsible and managed correctly. There needs to be good communication between the individual tasks. If any of these are missing then it is almost certain that the system will revert to a serial model and any concurrency is wasted. Concurrent models still rely on reaction to an issue. Project managers can build in regular meetings to facilitate communication but, in practice, it is still reactionary.

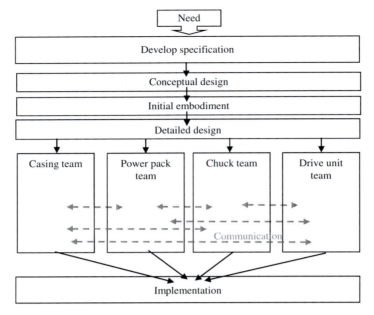

Figure 3.17
Concurrent model for power drill design.

For small projects the extra overhead generated by concurrent design projects, as illustrated in Figure 3.17, may exceed any savings generated. In larger projects, where individual elements are clearly visible, concurrent methodologies are regularly adopted. They are particularly prevalent in the automotive and aerospace industries.

However, one should not lose sight of the benefits of recognizing the iterative aspects of design. Nor should one forget that most designs have some form of subcontracting (e.g., packaging). It is here where concurrent models can help the small project.

3.3.1.4 Collaborative Models

Even concurrent models rely on one team informing another of an error in an aspect of the specification. The advent of the World Wide Web has made a new methodology more viable: *collaborative design*. Collaborative design is a model that depends on a repository being accessible, somewhere in cyberspace. All members of a design team have access to that repository. This changes the model, illustrated in Figure 3.18, slightly. It includes where the information is held: normally not a person but a secure web-based facility.

Modern Computer-Aided Design (CAD) systems almost universally have a built-in collaborative nature. Historically this exchange of information has been inhibited by the need for specialist software (that is often very expensive) to access data and to view designs. Modern web-based developments have made life much easier for collaborative models. One example is e-drawings®[12] , which is a package that allows everyone to see designs as they

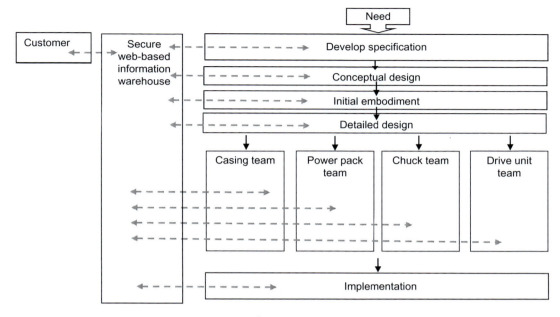

Figure 3.18
Collaborative design model.

develop without the need for specialist software. Note that this is the first model that formally brings the customer into the overall design process. Because access to data is now easier, it is inherently easier to give the customer access to ideas. In practice, this is often limited to the specification stage – after all we wouldn't want to teach our customers all of our tricks of the trade. However, this helps us to develop a good, robust specification and it makes it easier for the PDS to evolve as the project develops.

Communication is now between the data store and the teams. This communication can be *asynchronous* or *synchronous*. An example of synchronous communication would be one team in London working on one aspect of the design, and a second team in Oxford working on another aspect (a good example is a printed circuit board for the drive unit to go into the casing). Both teams can be working on the same data together using completely different computer packages; they can be talking to each other using Internet communication applications (such as SKYPE®) and looking at the results. This is all very practical in today's modern, electronic communication world.

However, there is a nasty thing that communications cannot sort out, and that is time zones. It is very easy to conduct synchronous collaborations between, say, San Diego and Los Angeles, but completely different when considering collaboration between London and Beijing – the

---

[12] For more information on the e-drawing environment go to http://www.edrawingsviewer.com/.

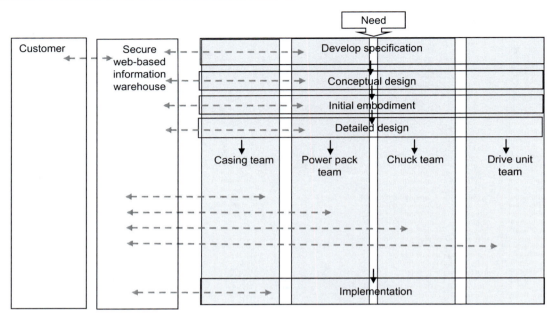

Figure 3.19
Holistic model.

time difference makes synchronous communication for a sensible period of time difficult. Here *asynchronous* collaboration becomes prevalent. All this model does is put a delay into the system. Using Figure 3.18 as the basis for the discussion, the casing team puts a suggested design into the data store; China begins office hours and the drive team passes comments and gives suggestions. The communication is not immediate – it is asynchronous.

Asynchronous communication will always be with us. But even office hours across the world can be changed to allow for synchronous work, if deemed necessary.

3.3.1.5 Holistic[13] Models

One last model merges the boundaries of a collaborative model. The collaborative model *still* allows one person to develop a specification in isolation. This is patently ludicrous. One person cannot fully understand the demands of a whole system and this inherently makes a good, strong, robust specification impossible to produce. It relies on people modifying the specification down the line – this is not good for a right first time ethos. Holistic models bring the teams higher up the development chain (Figure 3.19) and suggest that they, too, should be involved in the development of the specification.

---

[13] Holistic is from the Greek and means whole – or in our case include everyone. A search of holistic design on the web or in the library will result in sources discussing what color shirt you should wear, where to put that plant, or how to do feng shui. Let us recapture the term for the design fraternity.

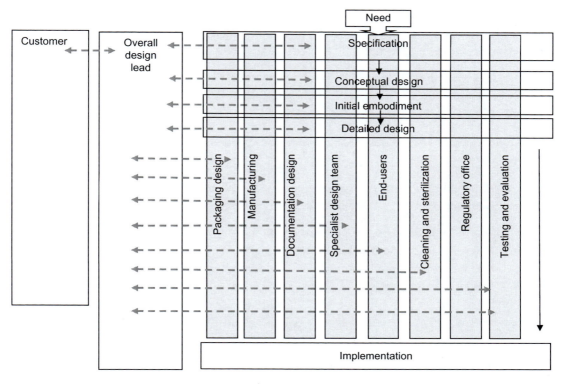

Figure 3.20
Generalized holistic model.

In a holistic model it is important that all potential partners are included from the start. This enables all of their experience, know-how, and detailed knowledge to come to bear to your design. Figure 3.20 illustrates how this may look for a typical medical device. Note that rather than being a top-down model, where the lead designer (or design lead) instructs the subordinates, this is much less dictatorial. It is not "bottom up" in approach – this just would not work. It is much more about "sharing the burden." The lead design team has the same overall authority, but they are much more concerned with ensuring collaborations occur in the way they should in order for the overall project to become a success. Indeed the design lead may not, actually, design anything! After all, do you really think that the project manager for the NASA space shuttle actually designed the shuttle itself? What this model ensures is that the drive and vision of the design lead sees the project through to completion.

3.3.1.6 Which Model Is Best for Me?
The previous models have no particular order of merit. They have been presented to indicate to you the importance of certain phases in the design process.

Firstly, it is important that the specification for the new product/device/piece of software is robust. It must be developed as a whole, and not by a single individual sitting alone in a darkened office. It is far too important to treat without due consideration.

Secondly, the expansion of the project by the generation of ideas is critical. The project must avoid "sacred cows." No stone should be left unturned; all ideas are valid until proven otherwise.

Thirdly, the reduction of the design space to a single potential solution is critical. This process must be robust in itself.

Fourth, the detailed design of the new product must, in itself, be robust. And the evaluation that the design meets the requirements laid down by the specification is paramount.

Only when these steps have been employed do we have robust design control. However, it is important that the lessons of the serial, collaborative, concurrent, and holistic models are learned. Try to be proactive and not reactive. Build iterative loops into your design to enable your design to change as it develops. Make sure you use all the modern communication tools to communicate with any of your potential stakeholders. And, finally, bring people into the design process as early as possible – this will make your life a lot easier in the long run.

3.4 Cross-Reference with Regulatory Requirements

As stated earlier, there are four main documents we must be sure we comply with. The first two are the FDA CFR 21 and 93/42/EC (the two main regulatory documents, as presented earlier). However their implementation is also covered by ISO 13485 and the FDA's *Design Control Guidance for Medical Device Manufacturers* (note: the FDA refers to ISO 9001). In essence, with respect to design, they all say the same thing – *do the right things and ensure you do them properly*. To paraphrase, they all state that we must listen to the customer and end-user, and produce a specification. They all state that we need to perform a robust design analysis, and they all ask for a final document (the technical file or design folder) that describes the final design and how it was arrived at. Up to now we have discussed models of design; now we need to leave the world of models and start to implement. At the same time we need to make sure that our design activities cross-reference with the relevant regulatory document(s).

As an example, I will cross-reference with the FDA document *Design Control Guidance for Medical Device Manufacturers* (FDA, 1997). Figure 3.21 takes the divergent–convergent model presented earlier and maps this to the essential sections described in the FDA guidelines.

Figure 3.21 demonstrates that internally we can use whatever language we like. However, when it comes to the regulatory authorities we must understand their requirements and their language. All that we do must map onto their framework. It is pointless to base your model on

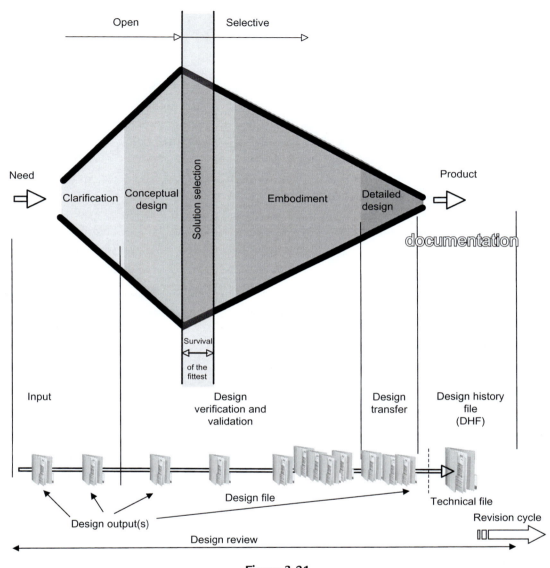

Figure 3.21
Divergent–convergent model mapped onto FDA guidelines.

the FDA guidelines or the EC guidelines as they use different languages. However, both refer to ISO 9001 clause 4.4 (and subsequently its equivalent in ISO 13485). Better to use this as the basis for your paper trail.

You will also note a similar example in the name of the design file. Once again, so long as you cross-reference your design file, your technical file, or your design history file they will all work. There is one simple reason for this: if you follow a structured design approach then it is only semantics and not content which is different.

It is worth summarizing what is meant by these terms. ISO 13485:2003 Section 7.3 concerns design and development of medical devices, specifically:

> 7.3.1 – Design and Development Planning: all about stages and who does what.
> 7.3.2 – Design and Development Inputs: clarification phase, building a PDS.
> 7.3.3 – Design and Development Outputs: records and documentation.
> 7.3.4 – Design and Development Review: making sure you are doing the right things.
> 7.3.5 – Design and Development Verification: checking that you've done what you said you were going to do and that the design meets the inputs.
> 7.3.6 – Design and Development Validation: Does it do what you said it would? Evaluation of your design under controlled circumstances; clinical evaluations.
> 7.3.7 –Control of Design and Development Changes: if you make a change, at any time, keep proper records.

Now that you have seen the design model, nothing in this list should seem daunting. As each of the next chapters develops, we shall make sure that we cross-reference the activity versus these requirements.

Notice that the requirements relate to documentary evidence; they do not tell you how to *do* the design.

When an auditor arrives they need to make sure that you have covered these seven requirements: you need to *prove* to them that you have done so. The best way to do this is to lay down procedures and use them. Hence, that is what the next chapter is all about; we are going to develop procedures that not only help you to design a device but make sure you meet the regulatory requirements.

3.5 Summary

In this chapter we met several design models. We were introduced to Pahl and Beitz's model, and Pugh's model. We then explored the fundamental principles contained within them and we found that it was a divergent–convergent model starting with a fundamental understanding of the requirements.

This results in the development of a fully populated *product design specification*. We were shown that for it to be robust, it is essential to include all stakeholders in its development. We then saw that the generation of potential ideas and the reduction of the design space into a single champion is an essential aspect in making our design process robust. Taking this potential solution – this embodiment – and making it a reality is the detailed design stage and here we saw that each individual aspect of the design is in itself another divergent–convergent process (ensuring that the best outcome is achieved).

We were then introduced to a number of models that help us to manage the design process. We saw that we need to accommodate iterations in our design and that we should incorporate the thoughts of others as early as possible. We also saw that through the use of modern technology, we are able to work collaboratively with our stakeholders, thus making a holistic, inclusive approach highly viable. Or, to coin a phrase, you want your band to be playing the same tune as you – all the way through the concert.

Tasks

1) Make sure you get copies of:
    the FDA document *Design Control Guidance for Medical Manufacturers*
    ISO 13485
2) Read and digest the FDA guidelines and Section 7.3 of the standard.
3) Map Section 7.3 of ISO 13485 with the FDA guidelines.

References

Bicheno, J., & Catherwood, P. (2005). *Six sigma and the quality toolbox*. Buckingham: PICSIE Books.
FDA (1997). *Design control guidance for medical device manufacturers*. FDA.
ISO (2003). *BS EN ISO 13485:2003 Medical devices – Quality management systems – Requirements for regulatory purposes*.
ISO (2008). *ISO 9001:2008 Quality management systems*.
Pahl, G., Beitz, W., Feldhusen, J., & Grote, K. H. (2007). *Engineering design: A systematic approach*. London: Springer Verlag.
Pugh, S. (1990). *Total design: Integrated methods for successful product engineering*. Prentice Hall.
Schrwazenbach, J., & Gill, K. (1992). *System modeling and control*. Butterworth-Heinmann.
Wikipedia (2011). *Newton's method*. Cited 13.06.11.http://en.wikipedia.org/wiki/Newton's_method.

Further Reading

Bruce, M., & Bennant, J. (2002). *Design in business: Strategic innovation through design*. Harlow: Prentice Hall.
George, M. L., Maxey, J., Rowlands, D. T., & Upton, M. (2005). *The lean six sigma pocket toolbook*. McGraw Hill.
Hurst, K. (1999). *Engineering design principles*. London: Arnold Publishers.
Jones, T. (2002). *Innovating at the edge*. Oxford: Butterworth Heinmann.
Ulrich, K. T., & Eppinger, S. D. (2003). *Product design and development*. McGraw Hill.

## CHAPTER 4

Implementing Design Procedures

4.1 Introduction

This has been the hardest chapter to write. Procedures are very personal; there is not one that is ideal for everyone. While I have attempted to present them in a logical order, the placement of this chapter has been problematic. Which subject does one need to understand first – the how or the why? There is no good answer. Hence I have decided to first show you why procedures are important. But be aware that *your* procedures cannot be designed and implemented until you fully understand the whole design process and all it entails, i.e., everything that's contained in the rest of this book. I apologize if this chapter seems juxtaposed because of this, but bear with me and all will be right in the end.

The previous three chapters presented design to you as an idea and as a concept. In particular Chapter 3 illustrated the basic concept of applying an engineering design model to satisfy the regulatory requirements laid down by the regulatory bodies. However, as we found out, the models and the regulations do not tell you how to actually *do* it. The subsequent chapters will cover the implementation of the design models we have already met. This chapter, however, covers the starting point for all regulatory implementations and makes sure that you can demonstrate that you meet the regulations. The best way to do this, and in fact the only accepted way, is to lie down and follow procedures. Hence this chapter will use the FDA guidelines and ISO 13485 (and the ISO 9000 family) as the basis for the procedures. You cannot go further unless you have copies of these in hand.

Remember that the medical device regulations demand one thing – to undertake your design activities correctly. The aim of the procedures is to ensure that the word "your" in the previous sentence is not just "you" but everyone with any design influence on the device. In other words, going back to the definitions in Chapter 1, everyone associated with the *manufacturer*.

Although this chapter attempts to present procedures and how to implement them, it does *not* present how your procedures should look. Procedures are as individual as you are – they need to work for you. Hence this chapter describes what they should contain and how to design them to fit your needs. It does give examples, but by no means are these "gold standards" and they should not be treated as such. Rather, use this chapter as a stimulus for you to begin to think how your company functions and how to make it function better – that, after all, is what a quality system is all about!

Medical Device Design.
DOI: http://dx.doi.org/10.1016/B978-0-12-415822-1.00002-7

4.2 Review of Guidelines

Table 4.1 provides a précis of the "design" sections of the FDA guidelines (FDA,1997) and ISO 13485 (ISO,2003) and the ISO 9000 family (ISO,2007).

You should now see that the guidelines all point in the same direction. Your processes must meet the issues presented in Table 4.1, *demonstrably*. The most common and most acceptable way to do this is to have documented procedures. How these are presented is your preference. It is perfectly acceptable to have written procedures; equally, it is just as acceptable to have *flowchart*-based procedures. You need to decide which form best fits your aspirations. The following sections will not prescribe which method to use but are intended to give you some ideas on how to formulate your procedures.

4.3 Overall Procedure

To fulfill the requirements for FDA Section B and its ISO equivalents you will need to formulate an overall design and development procedure. This procedure maps the route from input to output and how these interact with your company's other procedures (procurement for example). Figure 4.1 illustrates a typical flowchart for an overall design procedure.

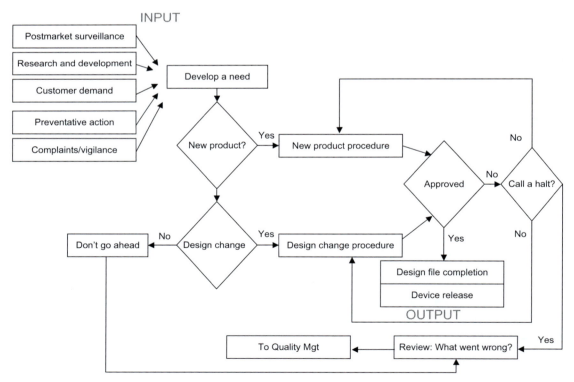

Figure 4.1
A typical input-related design procedure.

In reality, as Figure 4.1 illustrates, there are only three outcomes from an input: two outcomes are a brand new device or a design modification/change. The third option, the trivial solution, is that you decide not to follow up the need and abort the project. Hidden within this decision-making process will be a *risk analysis*; risk analysis is of paramount importance to the medical devices designer and though it may not formally appear in a procedure it should be assumed that it is undertaken.

It is important to see that the design process is input driven, as highlighted in Table 4.1. Where the inputs come from are up to you, but there are some that you have to cover. The first is *postmarket surveillance*. Here your whole company is listening to your customers, your specialist area, and scientists in your area. You will have a procedure in place that enables your company to distill all of the information that comes in and will produce inputs into your design process. The other main area you have to cover is *vigilance* (or *complaints*). Again, as a medical device manufacturer, you are required to have this procedure in place. As a design input this is called *preventative action* and, again, it needs to be "procedurized," but it is one aspect that, hopefully, never gets utilized. The ultimate sanction is a product recall – and we all can imagine the ramifications of that scenario! But on a positive note, logging and analyzing complaints can result in *design improvements* and hence this will almost always lead to a *design change*. One source of information you cannot afford to miss is communication with the customer (the end-user); your links to your sales force are so important. Do not fall into the trap of your sales team keeping information to themselves in a "they are my contacts" way; you must use your sales team to get as much market intelligence from the customer as possible. Often this will lead to design leads that you could never have imagined. You should note that we have effectively covered the types of need we discussed in previous chapters.

The next step is to develop a *statement of need*. This document should be approved and signed off on the basis that it is a strategic decision. Do we want to go this way? Do we want to make this change? Is it worth doing it? These are all the questions that you need to address. Six Sigma uses the "*5-Whys*" (George et al., 2005): if you ask "Why?" five times you will, nearly always, get to the answer you are looking for. In essence this is the opportunity for the first risk assessment of your design. As stated earlier, once approved the design process can only go two ways: it will either be a new product or it will be a design change. If it is not approved it will come to a halt.

From here, you will see that two new procedures are required: a new product procedure and a design modification procedure (we will explore these later). Both of these procedures will incorporate the majority of the design process discussed in the previous chapter. Both will result in an output: a design. However, the regulations state that output must be measured against input. The two procedures will do this but it needs signing off. Hence there is a final approval. Here we check that what should have been done has been done. Only then can the device go into full production mode. If it fails there are two tracks: either the design

Table 4.1: Précis of FDA Guidelines and ISO 13485

Row Number	FDA Design Control Guidance	ISO 13485	Synopsis	Section in This Text
#1	FDA 21 CFR 820.30	Section 7 Product Realization	The standards lay down the requirements that all medical device manufacturers design, develop, and ensure continual improvement of their devices.	4.3
#2	Section B Design Planning	7.1 Planning of Product Realization	All manufacturers of medical devices must have procedures in place to ensure that their devices are designed and developed correctly. All tasks should be planned.	4.3
#3	Section C Design Input	7.2.1 Determination of requirements related to the product	One of the main aspects of control is to ensure that the requirements and product specifications are clearly identified. Again, procedures to ensure it is done correctly are required.	4.3
#4	Section C Design Input	7.2.2 Review of requirements related to the product	See #3	4.3
#5		7.2.3 Customer Communication	See #3	4.3
#6	Section B Design Planning	7.3.1 Design and Development Planning	See #2	4.3
#7	Section C Design Input	7.3.2 Design and Development Inputs	See #3	4.3
#8	Section D Design Output	7.3.3 Design and Development Outputs	All design outputs are reviewed before release.	4.3
#9	Section E Design Review	7.3.4 Design and Development Review	Design processes(etc.) are reviewed at planned and strategic times.	4.4
#10	Section F Design Verification	7.3.5 Design and Development Verification	Checking that the design output actually meets the requirements stipulated in the design input!	4.5.3 4.4
#11	Section G Design Validation	7.3.6 Design and Development Validation	Checking that the design output is fit for its purpose within the field of intended use. This may include clinical evaluations or it may be checking that a "large" device works once installed.	4.5.3
#12	Section H Design Transfer	4.2.4 Control of Records	A very general paragraph stating the requirements to maintain controlled records, and the duration of retention.	4.5.5

(Continued)

Table 4.1: Précis of FDA Guidelines and ISO 13485 (Continued)

Row Number	FDA Design Control Guidance	ISO 13485	Synopsis	Section in This Text
#13	Section I Design Changes	7.3.7 Control of Design and Development Changes	As designs progress things change; there is a requirement to keep track of changes and the reasons for change. There is also a requirement to keep "old documents" (see 4.2.4).	4.5.4
#14	Section J Design History File	4.2.3 & 4.2.4 Control of Records	As per 4.2.4 but specific to the design. A clear record of a design and clear up-to-date description of the design (for manufacture, etc.) must be kept.	4.3 4.5.5

needs some further work, or (and this is by far the worst outcome) it is called to a halt and abandoned. In both cases the reasons why it has failed (the *root cause*) should be investigated and fully documented/reported so as to inform others. The failure could be highly laudable, but it could be something inherently wrong in your processes. You and your company should build a strategy of learning from "failures": "from the ashes of disaster grow the roses of success."[1]

4.4 Audit /Review Procedure

Two important aspects are hiding within the design review requirements. The first is the need to undertake planned design activities and hence formal reviews of the *design process* are required, e.g., weekly project meetings (we will address this in more detail later in this book). However the one thing most people forget is the requirement to actually ensure the design process is working and that procedures are being followed and documented. We will look at the former in later sections; for now we will examine the review process in more detail.

The reason for concentrating on this is that a medical device company need *not* be ISO 13485 or ISO 9000 certified to actually be a medical device company. It is not a statutory requirement (except in Canada). A non-ISO registered company will not have a sense of the importance of having this review procedure in place. But this does not mean that because they don't that they are exempt – far from it; it is very important to set *all* procedures and put them into motion. Hence, even if you are not ISO 13485 registered you should still try to work to ISO 13485 standards – it is not a great challenge.

One example of an important procedure is one that ensures that a review of the design process, activities, and outputs happens on a regular basis. All quality systems have an

---

[1] From the film *Chitty Chitty Bang Bang*.

auditing procedure to ensure that what is supposed to happen does happen; that any misgivings or failures are not "swept under the carpet" but are fully investigated to find the root cause; and that the systems and procedures meet the current requirements. No rules state how often this has to be done but it should be at least annually and the results have to be reported, formally, to a management meeting (normally to a *Quality Management Board*). It is probably sensible to audit procedures more regularly otherwise things can easily start to go wrong before anyone notices. Hence it is a good idea to have informal design reviews on a regular basis, say, bimonthly. Figure 4.2 attempts to show how this could work.

The main item Figure 4.2 demonstrates is that the review/audit procedure is continual. It is a part of the *continuous quality improvement* cycle. It attempts to show that you should plan a number of design reviews in your annual calendar, and that this calendar should culminate in an overall annual audit of your design procedures. These reviews do not replace your regular design meetings that go with each project; they stand above these and have a view over all of the projects and look at how they are functioning. The reviews are targeted at detecting any areas of concern and, equally as important, any areas of good practice.

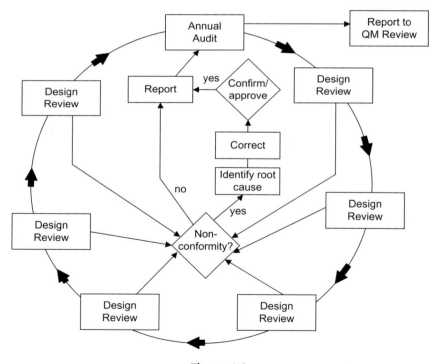

Figure 4.2
A suggested design review and audit program.

For example, a design review highlights that one person keeps forgetting to update revision numbers on part drawings (hence no one knows which drawing is the most recent). This is clearly an issue, and one that cannot wait for the end of the year to be resolved. The issue is raised as a *nonconformity* (this means it does not conform to the procedures) and a plan to rectify it is developed. The plan is implemented and then the outcome evaluated. Hopefully the person now updates the revision numbers and this is checked out. When confirmed that revision numbers are now updated the NC (nonconformity) is closed. A report is written, signed off, and submitted for the annual audit.

You can see that this review process makes sure that any issues are detected early and sorted out using planning, implementation, and final checks working as it should. But more importantly, it involves learning from a mistake and improving your design procedures to ensure it doesn't happen again.

The annual audit, however, has a more strategic quality role and looks at the design procedures as a whole: Are they working? Are there recurrent issues? Are there areas for improvement? By looking at the year as a whole, a bigger picture is formed. Also this annual audit provides evidence to the Quality Management team (and any external auditors) that the design control, as stipulated in the FDA guidelines and ISO 13485, is being implemented. It also provides the overall evidence required by the annual external auditors who will come to check that you can keep your medical device manufacturer status.

This can seem very onerous, and indeed some people do take the audit trail to the extreme! It need not be so if you remember what the audit and review processes are for:

1. To provide evidence that your design team(s) is following the procedures that you have set in order for your devices to be designed to meet FDA and EC regulations; and to ensure the necessary documentary evidence exists.
2. To enable your design management team to continually improve your design quality.
3. To identify any nonconformity issues and correct these as soon as possible, and before they have the chance to do any long-term damage.
4. It does not mean filling out multiple forms in triplicate!

It is not necessary to have graphical procedures. Personally I like them as I think they provide a flow of actions. Some people prefer to write documents – it is perfectly acceptable to write a procedure. We can reformulate Figure 4.1 into a document: Table 4.2 for example.

Table 4.2 suggests that each team gets a formal audit every year, but not all at once. You could chose to audit everyone at once, but this has one major failing – you need lots of internal auditors! One person cannot audit all 10 teams at the same time. Also, this enables the team leader from another team to act as the *internal auditor*; this gives the required "fresh pair of independent eyes" that enables the audit to spot what can easily be overlooked. Note, though,

Table 4.2: Suggested Annual Design Audit Procedure (for a Center with 10 Design Teams)

Month	Activity	Agenda	Participants	Evidence Produced
2	1st bimonthly design review	— Review previous NCs — Review design activity — Receive review report of 2 teams — Identify NCs — Produce action plan	All design team leaders	— Report — Action plan — Signed off NCs
4	2nd bimonthly review	— Review previous NCs — Review design activity — Receive review report of 2 teams/projects — Identify NCs — Produce action plan	All design team leaders	— Report — Action plan — Signed off NCs
6	3rd bimonthly review	— Review previous NCs — Review design activity — Receive review report of 2 teams — Identify NCs — Produce action plan	All design team leaders	— Report — Action plan — Signed off NCs
8	4th bimonthly review	— Review previous NCs — Review design activity — Receive review report of 2 teams — Identify NCs — Produce action plan	All design team leaders	— Report — Action plan — Signed off NCs
10	5th bimonthly review	— Review previous NCs — Review design activity — Receive review report of 2 teams — Identify NCs — Produce action plan	All design team leaders	— Report — Action plan — Signed off NCs
12	Annual design audit	— Review previous audit's NCs — Annual review of design activity — Bimonthly reviews — Identification of areas for improvement	All internal auditors	— Annual audit report — Action plan — Signed off NCs

that it is still important for the team leaders to keep things in control throughout the whole year and not just at audit time. The audits are not onerous; they may take two days and may be intense – but these are two very important days. The important thing to remember is that if they are planned well, are well structured, and are treated as constructive then the experience is pleasurable. If they are not planned well, are not well structured, and are initiated as combative (like gladiators in a coliseum) then they will fail and everyone will grow to hate them.

It does not matter whether your company is a "one-man band" or a multinational organization, the audit trail must be complied with. The only problem with a one-man company is who will be the auditor? Who will be the independent eyes? The obvious answer is to get someone else to do it, but remember that they have to know what they are doing – they cannot be an acquaintance from the tennis club.

Another important consideration is the individual audits/reviews themselves. You will need to develop an *audit plan*; this plan is best set by the annual design audit for the following year. The audit plan is, literally, a list of all of the procedural points the auditor must look at in order to confirm that there is evidence that the correct procedure has been followed. The auditor is not required to look at everything but selects randomly from the whole. Not every audit needs to cover every procedure, but by the end of the year *all* procedures must have been covered.

It is, therefore, obvious that the auditor requires training to perform his or her role correctly and diligently. Hence it is wholly appropriate for all internal auditors to go through an internal auditing training program.

Despite the above it should be noted that the auditing of your design procedures will, in the main, be controlled by Quality Management. However, the audit process has to work so it should be "designed" to suit your particular needs and not just copied from somewhere else (or even worse imposed by someone on the outside who does not understand design).

4.5 The Design Process

At last we come to the nub of the problem. We need to develop a procedure that describes how, as a company, we undertake our design activity. Unlike most procedures this is one that is hard to write down; it is more suited to a graphical flowchart approach. However, it is worth recalling that in Chapter 3 we described the design process in theory; all we need to do is put this theory into practice. The first phase was all about understanding the actual problem (*clarification phase*) and ultimately ending in a *product specification*. Hence it is worth building an overall model that starts here and then looks at each in turn. Remember, from Section 4.3 we will have two overall design processes: "New Product" and "Design Change."

4.5.1 New Product Procedure

Figure 4.3 illustrates a typical sequential design procedure.

You will notice that everything relies upon the identification of the need produced in the overall procedure and as shown in Figure 4.1. Firstly, a *project champion* or *project lead* needs to be appointed. It is this person's job to make sure that the project runs to schedule that it follows all the procedures, and that the document trail is complete. The procedure now expands the need by developing a full product specification in the *clarification procedure*. The procedures follow in line until final approval for release (we will examine these individual procedures in the next sections). In a documented procedure it is difficult to present anything but a serial, waterfall type activity flow. However the procedure only shows activity, it does not show information flow. Remember, the improved design models were all concerned with communication; the order in which the activities happened remained the same.

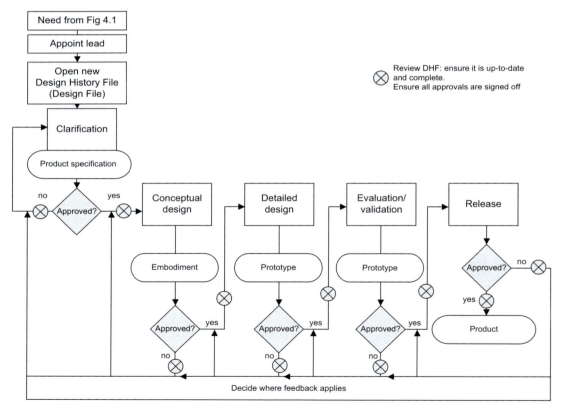

Figure 4.3
A typical new product procedure.

Note that after each procedural step there is an opportunity to review the outputs (Table 4.1, Row #8: Design Outputs) and confirm whether they are appropriate, correct, and meet expectations. Clearly, if all is well they are approved; this is done formally with a signature and date. A good way of doing this is to have a single product/project approval form with all sign offs listed sequentially. Also, it is worth taking the opportunity to review the design history file after each procedure. This enables you to see if anything has been missed or if any procedures have been incorrectly applied. It is far more efficient to keep this file up to date as you progress along the design path, to try and assemble it at the end – this way madness lies.

Furthermore, this is the opportunity to review any risk analysis – the risk analysis for the design is alive and changes as the design develops. It is, therefore, good practice to undertake a risk analysis review (Section 4.5.6) as a part of each approval stage as this will inform any feedback.

For example, your design for a clinical thermometer may not quite meet a requirement to measure temperature up to 42°C; it actually measures up to 39.95°C. According to the criteria this is a failed design and must be rejected. However, you conduct a risk analysis that actually suggests this is acceptable and poses no risk, so the design can go forward.

Alternatively you may have an automated insulin injection system that should not be able to overdose a patient, but under certain circumstances it provides a 110% dose. Here the risk analysis states the risk is unacceptable and hence the design is rejected with the feedback attached.

Hence a good risk analysis is a very valuable tool for the project leader! If all is not well, the noncompliance, and its *root cause*, needs to be passed back to the appropriate source. Identification of the root cause is very important – the risk analysis helps with this as does the "5 Whys." Note that even though the design may be rejected, the DHF is still reviewed because we still need to ensure that all has been done properly. Equally, as stated before, we need to learn from failures and be seen to be doing so.

After each of the individual subprocedures has been followed, a completed final product should be ready for release. At this point the Design History File/Design File/Technical File is closed as a new product. It now becomes the "bible" for that device and is maintained with the utmost care. However, as we shall see later, it will be reopened on a regular basis as the effects of *postmarket surveillance* kick in.

4.5.2 Clarification/Product Specification Procedure

This procedure is important as it meets all of the requirements associated with input in the FDA guidelines and in ISO 13485. This is, by far, the hardest of all the procedures to develop as the potential inputs are infinite. However, I have tried to summarize the sources in Figure 4.4 – this is the bare bones of such a procedure.

Figure 4.4 illustrates the process to develop the full product specification. There are numerous influences on the product specification and, much like a cloud, they tend to hover above

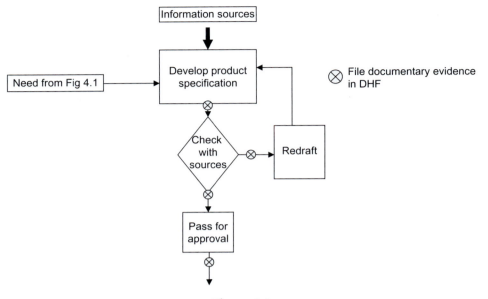

Figure 4.4
A typical product specification procedure.

the process, influencing the process but always just out of reach. It is the designer's task to identify the elements that are influential and bring these into being. Some of the sources will always have influence (standards for example); some will not (trade literature for example). But none can be excluded without good reason. It is difficult to do anything but list all of the sources you intend to use. A very important source for the specification will be an *initial risk analysis*. This initial analysis will help you to understand the whole of the area in which you will be designing. Understanding the risk is a large step towards understanding the reality of a situation.

The ultimate aim of this procedure is to demonstrate the communication between the product specification developer and the sources. This procedure must have built-in continuous feedback to enable the primary sources, i.e., the end-users, patients, and customers, to have a significant impact on the specification itself. This will enable you to develop a highly robust specification. Note that each step produces a draft for comment, and each iteration does as well. It is important that this is well documented and kept within your design history. Once the team is happy with the product specification it can be passed on for final approval before the next stage begins. Remember that the input is the statement of need and the data from the sources; hence the product specification (the output) must meet the requirements of the need and reflect the requirements of the sources.

Please also note that the specification should not only relate to the device itself, but also any supporting documentation, etc. For example all devices will need labeling and the specification should address this need. A further example is that the device will need instructions for use; the specification should address this too. It is too late to consider this at the end when the device has been made. It is obvious that the holistic model best suits this phase – *talk to everyone*!

4.5.3 Detailed Design Procedure

It is noteworthy that this procedure is important to the realization of a design but there is no direct section in either FDA or ISO for that relates to it. However it is argued that one is required to demonstrate meeting the requirements of Table 4.1 Row #2: Design Planning. In relation to the overall procedure it is easier to combine the creative phase and the detailed design phase into one then to split them. Although they work as two separate phases, to show them as two separate procedures makes little sense as they are so entangled.

It is important to note that the first phase (creative phase) is intended to expand the design space and then condense it to one embodiment. The actual methodology will be examined in more detail in subsequent chapters. The subsequent phase (detailed design phase) takes this concept and "makes it real."

It should be noted that Figure 4.5 could be used for any of the embodiment phases of design. It could be used for the first prototype; it could also be used for the final design for manufacture; it could even be used to select an evaluation method. The *same* overall process applies. As with previous procedures risk analysis at each approval stage is of great importance. However this procedure also writes it into the specific design activities; as each of them will result in judgments being made, a risk analysis helps to justify the judgments.

A further noteworthy item is that this procedure starts with the setting up of a team. This means selecting the appropriate members (both internal and external); it also means the setting up of milestones and timescales.

4.5.4 Design Verification/Validation/Evaluation Procedure

To meet the requirements in Table 4.1 Rows #13 & 14, the design needs verifying and validating. Verifying is concerned with making sure the outputs meet the inputs. Validation is concerned with checking that the output works within a clinical environment. Both are very similar in concept hence one procedure could be used to form the basis for both; I have called this *design evaluation* (Figure 4.6).

The important aspect here is the fact that the device will need validating or verifying against some criteria. These criteria need to be selected and the evaluation protocol must be devised and approved. During the design process this procedure will be used and tested lots of times, as it is the basis of checking appropriateness of a design. It should be noted that clinical trials, etc. come under this procedure. It is also worth noting that the final evaluation, verification, and validation will have been specified in the initial product design specification.

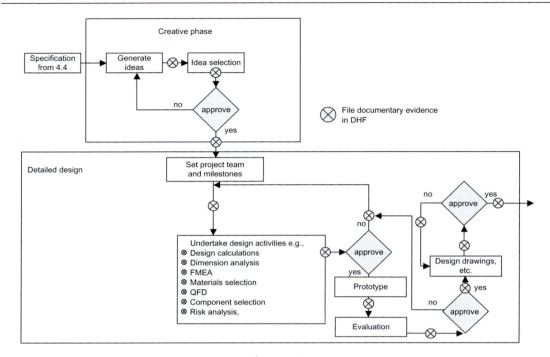

Figure 4.5
A typical detailed design procedure.

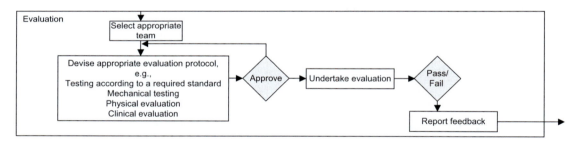

Figure 4.6
Design evaluation procedure.

4.5.5 Design Changes

This procedure is required to meet Section I and ISO 13485 7.3.7. In essence it has two main purposes: firstly to ensure that any changes are made for the right reasons and undertaken correctly; and secondly to ensure that the changes are made obvious so there can be no mistake that a change has been made.

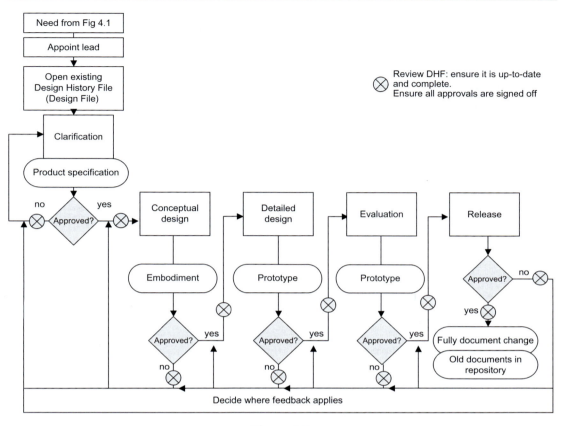

Figure 4.7
Design change procedure.

It should come as no surprise that a design change is similar to a full new product design. There are, however, two main differences. The first is that this procedure will always relate to an existing design file, hence the first step is to open an existing file. What follows is the same. All that has been described before still applies. More often than not, it is a small change that causes a company a lot of grief. So do not fall into the trap of thinking that this procedure is only about documentation – it is about making a change properly. Hence, having adopted the change, make sure you assess the risks associated with said change. Finally, the end of the procedure is different; here one ensures that the change (or changes) is fully documented in the file and the old documents are stored in a repository (Figure 4.7).

It is important to note that some design changes can be monitored internally; for others you will be obliged to inform the regulatory bodies. We will be looking at this in more detail later.

4.5.6 Control of Documents

It is very important to have a *document control*. This can be written but it must cover:

1. How long documents are kept
2. Who is responsible for maintaining the design history file
3. Where the original design history file is kept

But as you have already seen, the procedures force you to control your documents.

4.5.7 Risk Assessment Procedure

Throughout this chapter the term *risk assessment* has been forced into your subconscious. This is for a very good reason. Conducting a risk assessment when you make a decision is good practice. It forces you to inspect the ramifications of your decisions. Suppose, for example, you decide to change a single component in a large device. The change may be innocuous, say, reducing the diameter of a pin that holds a screen in place. However you may have overlooked that this change now makes all of the previous devices sold different – how will you ensure that if someone asks for a replacement pin that they will get the right size? What is the risk (or risks) if they get the wrong size? You would be amazed to find how many devices have suffered from ignoring this simple analytical step. As it is so important it has its own standard – ISO 14971: Application of risk management to medical devices.

The following procedure (Figure 4.8) is general. It must be adapted to meet the specific circumstance in which it is to be used, however it is a usable procedure; we will examine how to perform a risk analysis in more detail later.

The important part of this procedure is to have an approved pro forma to complete. Without this the whole procedure will fail. As with previous procedures it is important to document the analysis at key stages (and, of course, file). It is also important to check that any suggested actions from the risk analysis are actually undertaken, reported, and filed.

4.6 Implementing a Procedure

While a flowchart or a document describes how a procedure works it is not a formal procedure until a number of things happen. The first is that it should be presented correctly. The procedure will need a title, it will need a version number, it will need to be signed off and dated, and finally it will need a table recording changes to the procedure. The last thing is that during signing off it will need to be formally included in the company's *quality manual*. Figure 4.9 illustrates a typical complete procedure.

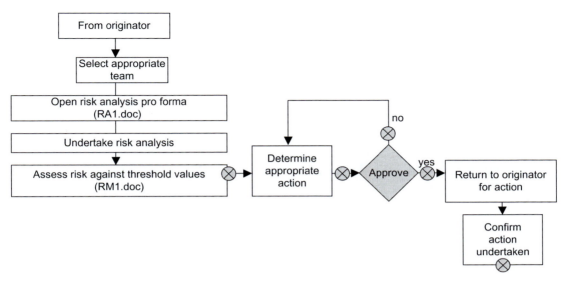

Figure 4.8
Suggested risk analysis procedure.

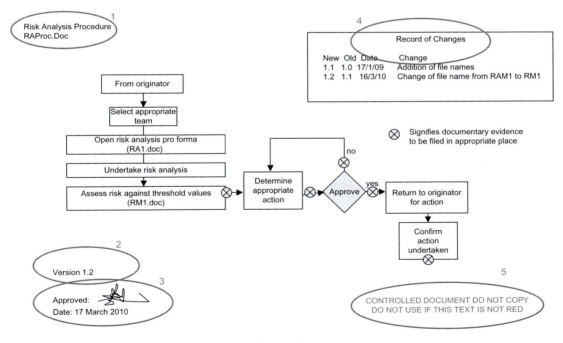

Figure 4.9
Typical procedure layout.

The rings would not exist on the real thing – they are there to help me clarify things:

- Item 1 is the title of the procedure and its file name. The inclusion of a file name is important in this electronic world.
- Item 2 is the version number. It is worthwhile having a sheet in the quality manual, and on a wall, that gives the most up-to-date version numbers of all procedures and controlled documents.
- Item 3 is the proof of sign off.
- Item 4 is the record of changes. As each new version evolves something will change, and this table enables that to be tracked. Obviously the table will grow and there is no need to have all changes listed (or your procedure will be a small picture in a large table of changes) but at least the last two/three changes should be logged. As the old procedures are placed in your repository, this historical record will develop.
- Item 5 is optional but is good practice. This statement should be produced by a red-ink stamp that is held by one person only. In that respect they are the only people who are able to print off this document. If someone tries to photocopy it the red text comes out black – hence it is obvious that it is a copy. However, modern IT has superseded this and most companies now have a secure FTP server where up-to-date (and only the up-to-date) versions are kept; everyone has access and there is no excuse to not have the latest version.

4.7 Summary

In this chapter we met procedures. We saw how the procedures are used to demonstrate meeting the requirements for design control in FDA CFR 21 and ISO 13485 and how they are used to ensure that devices are designed right the first time – every time. We further saw that the procedures ensure that all of the design documentation is controlled. While there has been an attempt to provide procedures that are general in nature it is important for you to develop your own procedures as this helps you to fully understand your company, your product line, and your customers.

References

George, M. L., Maxey, J., Rowlands, D. T., & Upton, M. (2005). *The lean six sigma pocket toolbook*. McGraw Hill.

FDA (1997). *Design control guidance for medical device manufacturers*. FDA.

ISO (2003). *BS EN ISO 13485:2003 Medical devices – Quality management systems – Requirements for regulatory purposes*.

ISO (2007). *ISO 14971:2007 Application of risk management to medical devices*.

## CHAPTER 5

Developing Your Product Design Specification

5.1 Introduction

We have already looked at the importance of a good specification. It is the cornerstone of any good design. In this chapter we shall take a look at the basics of a product design specification (hereafter called a PDS). First we shall examine how to develop a good statement of need; subsequently we shall expand this to a full-blown PDS.

I cannot stress enough how important the PDS is. It is the essential *input* to meet FDA, Medical Device Directive (MDD), and ISO 13485 requirements. But, and more importantly, it is your only weapon in the battle between you and your customers. Too often the customer forgets what they asked for, doesn't understand what they really want, and just keeps changing their minds. The PDS is your weapon that enables you to say "this is what you agreed to last time." More importantly, and as we have seen earlier, the more effort you put into understanding what is "really needed" the easier everything else becomes.

A good rule of thumb for a PDS is that you should be able to give a completed PDS to another designer (of equivalent skills) and without any further conversation they should be able to understand what is needed, fully. That is our task for this chapter: to be able to write a fully populated PDS.

Before we start, there's one last thing to note. Too often a PDS is confused with the "specification" one finds in sales literature. They are not the same thing. In sales the specification describes the characteristics of a product after it has been designed, and customers use it to discern between competing products. Do not use these as the basis for your PDS!

5.2 Developing the Statement of Need (or Brief)

As described in previous sections, the starting point of the design process is the identification of a need. This need has to be articulated and approved before any further effort (and hence costs) are attributed to the project. We have met the types of need, and it is immaterial where the seed comes from – how we define the need is the same.

Medical Device Design.
DOI: http://dx.doi.org/10.1016/B978-0-12-415822-1.00002-7

5.2.1 Identifying the "One Thing"

If any of you remember the film *City Slickers*[1] you will remember the famous line, and I paraphrase:

> *"You need to understand one thing; the hard part is to know what the one thing is."*

This is pertinent to any design. All designs have one main aim, one thing that makes them different to all other things, the one thing that makes them what they are. You need to articulate this "one thing."

For example, consider a passenger aircraft. What is the one thing? One could say it is a vehicle to transport passengers – but that could also be a bus, a car, a van, or a cruise liner. One could say it flies – but so do a bird, a kite, a jet fighter, and a hot air balloon. Quite clearly the one thing about a passenger aircraft is that it is "a vehicle to carry a number of passengers that flies." Now this could be a passenger aircraft, a zeppelin, a hot air balloon, a rocket or even the shuttle…that does not matter. They all meet the essential "one thing"; the design process sorts the rest out.

This is by no means easy. Do not be surprised if you find it challenging, difficult, and tiresome. But I promise it will be worth it in the end.

CASE STUDY 5.1 Define the "one thing" for a corkscrew

This is not a medical device I know, but probably one that is close to the hearts of most designers! What is the main objective of a corkscrew that makes it what it is? First forget the name corkscrew – this immediately produces the "sacred cow." Go back to basics: What has it got to do?

It is obvious – the main aim is to remove a cork from a bottle in one piece.

Note: By going back to basics and really examining the "one thing," sacred cows have been banished.

5.2.2 Formalizing the Statement of Need

The trouble with "the one thing" is that it ignores all the other things that need to be done too. So while it is very useful to get to the heart of the matter, it does not fully state the need. Equally, if we go too far into the subject we start to write a PDS and not a brief on which to make a commercial decision. Ah, now there's the rub…we need to make a commercial decision on whether to go forward or not. Hence the *statement of need*, or *design brief*, is a commercially led document. It asks the questions

- Can we do it?
- Can we afford to do it?

---

[1] *City Slickers*, MGM Studios, 1991 – a hit comedy in the early 1990s.

- Can we afford not to do it?
- Who wants it?
- How many want it?
- For how much?

…but not necessarily in that order. The design brief is also the start of the formal design process and hence needs formal approval. The best way to achieve this is to formulate a simple document, or pro forma, to complete. We are in a quality process so this document cannot be ad hoc: it needs to be developed, written, and approved before it can go ahead.

Figure 5.1 is an example of an approved statement of need pro forma. As with all other example documents in this text it is by no means an exemplar for direct copying, but is a basis from which to develop your own.

It is so simple to produce a pro forma on modern computers that there is little excuse not to have one. Note that this is a controlled document hence it has a unique name, a version, and formal approval. It should be housed in your company's quality manual.

It is worthwhile giving each new statement of need a unique project number – this makes cross-referencing so much easier. Hence it is worth keeping a log/track record of all statements of need and their outcome. The product title need not be the name the product will keep forever but can be a good secrecy ploy. In the First World War, when a new armored vehicle was being developed, the military didn't want its secrets to get out; hence it had lots of different companies making lots of different bits that when put together made this new vehicle. When the companies asked what they were making they were told it was a new vehicle for carrying much needed water to the troops, which was quite logical since it was a big metal box on wheels; hence everyone called it *a tank*…and this name has stuck ever since. If you are working on something brand new and secret don't give your project a name that gives it away. You don't want your competitors knowing what you are working on, so call it something that people can recall but that means nothing outside of a closed circle of friends.

We have covered the description of the need earlier. But this section must contain all the information required to make a reasoned commercial decision on whether or not to go ahead. The minimum should be the overall aim of the product (the one thing), where the demand has come from, potential market size, and potential sale price. It is also worth stating whether this is totally new to your company or not (e.g., a cardiology specialist going into diabetes management) and whether you have the expertise to do this. An accurate indication of Cost to Market (R&D) is essential.

The next section details the evidence submitted. These could be written demands from customers, copies of market research reports, or transcripts from focus groups. This section enables you to put all of that evidence in one place.

Medical Device Co Inc.

Statement of Need Pro forma

Project Number

Product Title

Description of Need

Evidence Submitted

Approved / Not Approved

Signed

Date

SoN.doc   version 1.0   Approved by:

Date: 17.5.2010

Figure 5.1

Example of a statement of need pro forma.

Note that although Figure 5.1 looks like one sheet of A4 it need not be. Clearly a project that costs the company £50 is not going to get as much scrutiny as one that is likely to cost $1 million. Hence one would expect much more evidence to be provided with large cost projects but this does not mean that smaller ones get no attention – lots of small useless projects cost as much as one big useless one!

The final part records approval or rejection. A single project may go through this process several times before finally being approved, so just throwing the form back is unhelpful. Clearly, if rejected the reasons why should be given. Equally, if accepted the reasons why should be given.

What happens next? If accepted the new product procedure or the design modification procedure kicks in (see Chapter 4). If rejected a comment is made in the log and a copy of the forms filed; originals plus comments are returned to the originator to decide whether to do some more work or stop. It is often the comments from the panel/board that make this decision for the originator.

5.3 The Product Design Specification (PDS)

Much has been said about the PDS, and as you will know it is thanks to Pugh (1990) that we have this useful tool. In the end it is immaterial whether you call it a PDS or a specification; just make sure the word "specification" appears for your regulatory trail.

While we are on the subject of regulations they are a very good starting point for your PDS. The EC directive (93/42/EC), for example, contains an annex called Essential and General Requirements; these are things that your device must meet to be classed as a medical device. The FDA and other bodies have similar sections. You should have a copy of these in hand and tick them off as you your PDS development progresses to ensure that all has been covered. Another good tip is to cross-reference your PDS against these requirements; this is easily done by numbering each item in your PDS and using this number as the basis for the cross reference.

We demonstrated (Chapter 4) that it is this document that highlights all of your inputs. Firstly I shall present the inputs in a logical manner to enable you to formulate a PDS; secondly I will show you how to get the information to fill it with. The latter, I assure you, is much harder than the former!

5.3.1 Essential Elements of a PDS

As you can imagine, to list the potential content of a full PDS would be both time-consuming and fruitless. It is more commonplace, and more beneficial to both you and me, to present the basic elements and let you complete it using your own specialism. I have already presented some texts for you to refer to. Developing your PDS is where you really do need to refer to the literature. Do not just rely on this chapter; not because I do not know what I am doing but because I cannot possibly cover every eventuality in the medical devices world.

In most textbooks on design, e.g., Hurst (1999) and Ulrich & Eppinger (2003), you will find that a general PDS has the following sections:

- Introduction and Scope: a resume of the need.
- Performance Requirements: a complete dialogue of what the "thing" needs to do and how to behave.
- Manufacturing Requirements: a complete dialogue of how the "thing" should be made, treated, packaged, etc.
- Acceptance Requirements: a complete dialogue of what needs to be done before the "thing" can be put on the market.
- Environmental Requirements: a dialogue concerning environmental impact, disposal, waste, etc.

These section titles have probably left you none the wiser. They are far too brief to help. I prefer to use the term "requirement" when something is *required*; the term "factor" when something puts limits on your design (closing down your design space); and the term "indicator" when dealing with *design objectives* that one would like to achieve. Hence I prefer to categorize by source (or if you wish "voice of the…"):

- Customer
- Regulatory and statutory
- Technical
- Performance
- Sales
- Manufacturing
- Packaging and transportation
- Environmental

You may well wish to add to the list – that is absolutely fine. Equally you may wish to expand some into smaller bits – that is fine too. There is no single PDS format to stick to. However, let us look at each source in turn and see what it is we are supposed to be discerning.

What is an actual PDS? It is a document and as a consequence it is a controlled document that you need to develop and approve beforehand (and put into your quality document). You will not be able to include everything, but you should include your main headings and numbering. Figure 5.2 gives you some idea of how one looks. It is always good to start with a summary of the statement of need to link it to the PDS; the two documents are linked and so go together as one. Remember that as your design develops you will need PDS documents for individual subassemblies and components; hence a summary links these PDS documents with the main one. Also, you will have started part numbering the main device and its subcomponents so this is accommodated too.

Medical Device Co Inc.

Product Design Specification

Originator	Date

Project Number / Part Number   Version:

Product Title

Summary

1. Customer:

2. Regulatory and statutory:

3. Technical:

4. Performance:

5. Sales:

6. Manufacturing:

7. Packaging and transportation:

8. Environmental:

Approved / Not Approved

Signed

Date

PDS.doc   version 1.0   Approved by:
                                               Date: 17.5.2010

Figure 5.2
Example PDS pro forma.

It is unlikely that a PDS will be one sheet of paper. It's far more likely to be at least 10 – the more complex the device, the more extensive the PDS.

There is no requirement to have sections. However, learning to write a PDS with sections is advisable in order to help you remember all that should be included. Later you may find that sectioning becomes less important than actually stating the source of the information.

5.3.1.1 Customer

All devices will have a customer; in modern parlance the *end-user*. It is often very difficult to find the actual end-user as the person that initiates the need may not actually be an end-user. Equally, the person who actually buys the device may not be an end-user (but is certainly a customer). This is where things such as workshops, focus groups, and talking at conferences really help. The voice of the customer has been overlooked on many occasions, and often to the detriment of the company concerned. However, this is the easiest section to complete as all the thinking is done by others – you act as a filter. All of the other sections in the PDS are wholly yours. Hence there is no reason why this section should be devoid of content. Furthermore, as the voice of the end-user becomes more prevalent in the regulatory framework (which it is), this section needs to be demonstrably visible.

Please, please, please do not think that just by talking to a surgeon you have discussed items with an end-user. They are often the last link in the chain; an important last link, but still the last link. Before them come purchasing: What is it they want to see? Has it got to be blue? Has it got to be cheaper or within 10% of an original price? What do the sterilization and cleaning staff want? Does it need cleaning trials before it can be accepted? Do they need holes in places for washing that you never thought of? What does the nursing staff want? (They will be fetching it and unpacking it.) Does it need to be below a certain size for the shelves? Does it need to be below a certain weight for them to carry? Would they like it to be a special color to stand out from the other items? It is important that with the comment you include the source of the comment. This is useful for "backtracking" (i.e., going back to the source to confirm) and for cross-referencing.

All of the above is concerned with finding out what the customer "actually" wants. As stated earlier, often they do not know what they want – you have to tease it from them.

I cannot stress enough the importance of the voice of the customer. It is only by getting them on board that you will really fully understand the problem you are trying to solve. Furthermore this section will be of great importance when we address the *House of Quality*.

Table 5.1 illustrates how different sections of the customer base see simple things like color differently. It is your job to filter these down to some form of consensus to enable a sensible PDS to be written.

5.3.1.2 Regulatory and Statutory

This is basically common sense. We are bound by rules set down by the FDA and by the European Commission – we have to meet those so make sure they are stated. The obvious one is:

> *It must meet the essential requirements as detailed in…*

However there are many more standards and regulatory requirements that your device will have to meet, some of which you may not have even thought of (those of you who have powered devices may come under numerous regulations from noise limitation to electromagnetic

Table 5.1: An Example of the Customer Section in a PDS

Section 1: Customer		
Number	**Comment**	**Source**
1.1	Color: Theater staff requested it not to be black as this is common and causes confusion between components of various companies.	OR theater staff
1.2	Color: Central sterilization requested that the color be resilient enough to cope with the latest washing regime. Lots of "older" devices tend to lose surface color using newer washing machines.	Central cleaning and sterilization staff
1.3	Color: Surgical staff really like the plain surfaces to be shiny, but not so shiny as to cause reflections from the OR theater lighting.	OR surgical staff

compatibility). The important thing is that you conduct a thorough review to find out which regulations, standards, and guidelines your device must adhere to. Don't forget that there are different standards for the same thing in certain countries so just doing something to a British Standard does not mean automatic correlation to an American ASTM standard. Standards may well be expensive items but, as we will see later, you have little excuse not to refer to them.

This is also the section where you may wish to include any requirements for instructions for use and languages. You will also need to include labeling requirements.

You should have made an estimate of the classification of your device; this may well change as your design progresses but the higher the classification the more rigorous the design process. Hence it is better to start at the right level than to try and increase your rigor part way through!

In Table 5.2 you will notice a comment called the standards review. It is worthwhile to conduct a review of standards to find which are applicable (and those that are not) and to write a brief document so that the specification can be brief but also supported by a report of depth.

5.3.1.3 Technical
As you learn more about the problem you will start to think of your own criteria – these are classed as "technical" (some people call this *functional requirements* but it matters not). For example, will the device's power supply be 110V or 240V? You will also start to understand any loading the device may be subjected to. This is your opportunity to use your experience to start to lay down the technical boundaries to the design space.

Table 5.2: An Example of the Regulatory Section in a PDS

Section 2: Regulatory and Statutory		
Number	**Comment**	**Source**
2.1	Medical Devices Directive: Must meet the essential and general requirements of the Medical Devices Directive.	EC/97/42
2.2	FDA: Labeling must meet FDA requirements.	FDA 21 CFR 801
2.3	Material: Material must comply with ISO 5838-1:1995.	FDA recognized consensus standards. Standards review.
2.4	Testing: Screws to be tested to meet ASTM 543.	Standards review.
2.5	N 60601-1: Medical Electrical Equipment –General requirements for basic safety and essential performance.	Standards review.

It is within this section that you describe the environment in which the device operates (Table 5.3). Will it be steam sterilized or gamma irradiated or both? Will people cover it with alcohol and set fire to it? Don't forget it has to be transported so although it may be used in a nice clean OR theater, it may have flown in a baggage hold at −20°C and then crossed a desert at +40°C. All possible technical limitations need to be considered. You need to use all of your experience and discussions with end-users, supply chain, and sales force to fully understand the technical requirements your device will be put through. You also need to think about protecting the user. Are you emitting noxious fumes or ionizing radiation? Will the user be sitting in front of a VDU screen all day? What does your design have to do to meet these technical limitations?

Don't forget ergonomics and "usability." At the end of the day someone has to use your device; hence they must be able to *use* it. Ergonomics, man-machine interface, usability and anthropometric data are all concerned with *fitting the device* to "man." Quite often this comes from the customer in the form of "I must be able to do this with my left hand"; you need to decipher this into technical data. What, technically, does "using with the left hand" mean?

As with the previous section, this section contains two further reports: investigation and focus group. We shall meet these in more detail later.

5.3.1.4 Performance
Now we are getting into interesting territory. Just what does the device have to do and how well should it do it? Imagine setting the performance for a car. How fast should it go? How fast should it accelerate? How many miles per gallon should it do? All of these are performance characteristics. You will need to set criteria that someone can use to assess your device. Again, some of this will be determined from discussions with end-users, however some will come through your study of the subject area. For example, your device may need to measure temperature from 0–100°F with an accuracy of ±2%. How many times can it be used

Table 5.3: An Example of the Technical Section in a PDS

Section 3: Technical		
Number	**Comment**	**Source**
3.1	Static weight: The maximum static load is based on 95% male in the UK = 96 kg.	World Health Organization
3.2	Dynamic loading: Due to ambulation maximum dynamic load is 120% body weight.	Introduction to Biomechanics**
3.3	Environmental humidity: Can be used in totally dry to totally immersed environments: hence humidity 0–100%.	Investigation report
3.4	Environment temperature: Can be used from North Pole to Equator: hence −40°C<T<40°C.	Investigation report
3.5	Working temperature: In sterilization temperature can reach +130°C.	Focus group report
3.6	Measure temperature: The device should measure temperature from 0–40°C.	Focus group report

**When quoting from textbooks or journal papers use Harvard referencing – most university websites have a free guide.

before it requires servicing or calibrating? This section will almost certainly be populated with numbers (Table 5.4).

Do not forget that when it gets to the end-user the device has to operate as it did when it left the factory. How is that to be judged and ensured? This is important for many devices as it is written in stone in all regulations that the device must perform, and be shown to perform, as intended in situ – not just in the factory or in the laboratory.

5.3.1.4.1 Biomechanics

It is within the "performance" and "technical" sections of the PDS that we should find information related to biomechanics. Biomechanics is the study of the animal as an electro-mechanical or mechatronic system in order to reveal distinctive parameters. I am unable to give this subject the justice it deserves; there are whole books on small aspects of the subject. So, I have decided to mention it by name, mention its importance, and tell you how important it is that you obtain access to relevant biomechanics reference texts. There are numerous books on biomechanics; some are all about modeling; some reveal actual numbers; some are about human movement; and some are about electrical modeling of neurological systems. Whatever you are designing there will be related information within a biomechanics textbook, somewhere. Two books you may wish to consider for your shelf are Enderle and Bronzino (2011) and Webster (2009). The first presents biomedical engineering concepts from first principles; the second concentrates on medical-related instrumentation.

Table 5.4: An example of the Performance Section in a PDS

Section 4: Performance		
Number	**Comment**	**Source**
4.1	Temperature measurement: Measure T from 0–40°C to an accuracy of 1% full scale deflection.	Focus group report
4.2	Deflection: The device should deflect 1 mm under body weight.	Focus group report
4.3	Operating system: The software should operate successfully on both PC and Mac operating systems.	Focus group report
4.4	Servicing: Should operate successfully for 25 uses between servicing intervals.	Investigation report

---

CASE STUDY 5.2

What dynamic loading would you expect for an average male walking in normal footwear?

From a standard gait analysis graph (Figure 5.3) peak loading is about 1.2 × body weight.

Note: You do not need to be an expert in biomechanics to be able to design medical devices, but you should be an expert in referring to the relevant texts!

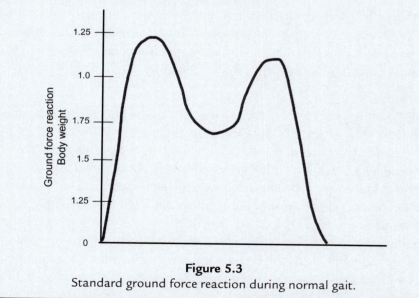

Figure 5.3
Standard ground force reaction during normal gait.

5.3.1.5 Sales

Your sales force will always have an input. They know the marketplace, they know the buyers, and they know the problems associated with selling the devices. Get them on board straight away and ask them what it is that will help them to sell your device. This may be the one chance you get to plan tests that are not just for meeting regulatory requirements but are actually there to produce marketing collateral before the device goes out for sale.

Your sales force will tend to bring things into the design with the intention of making the device "sellable" (Table 5.5). They will, as with customers, pose questions rather than requirements. For example, can this be done? Can this be achieved? One of the most important facts is the selling price (with, of course, profit margin)…this tells you what your budget is!

5.3.1.6 Manufacturing

This section has two aspects. Manufacturing can set limitations, for example you may be unable to manufacture in composites; there can also be requirements for production, for example you may have to make this device in a *clean room*.

More importantly, this is your chance to get input from the personnel who will actually make the "thing." Too often things get to the shop floor only to be returned labeled "unmakable." The manufacturing team at your disposal is a great asset; bring them in at the start to look at how manufacturing capability not only limits your design but also to possibly open your eyes to something you never thought of (by and large it is the latter which is the most invigorating).

You may have to look at installation too. Will your device have to be installed (this is a part of the manufacturing process)? Will it require calibration, setting, or adjustment in situ? Who will do this? What is required to do this? Will your device need assembly – does it arrive like flat packed furniture?

One of the main reasons for estimating the classification is for this section. The higher the classification the greater the level of importance laid on your manufacturing facilities. It is quite apparent that the cleanliness of a manufacturing facility for a syringe is far more rigorous compared with that for a chair for a hospital waiting room. The classification of your device gives you an indication of the rigor required.

Quite often a company will have internal rules, sometimes to minimize stock on the shelves and sometimes just from a bad experience with a previous product. Sometimes they should be adhered to, sometimes they can be questioned – but they should not be ignored. You may also find that your suppliers do not like to mix materials. Many implant

Table 5.5: An Example of the Sales Section in a PDS

Section 5: Sales		
Number	**Comment**	**Source**
5.1	Selling price: Not to exceed £50 with a gain margin of 60%.	Sales report
5.2	Color: Purple is going to be the "in color" at the time of sale.	Sales report: Market trend report
5.3	Operating system: Apart from mains electricity the use of "solar power" would give a USP.	Sales report
5.4	Color: Could any knobs be individually colored for ease of training?	Sales report
5.5	Sales: Estimated demand is 400 per year.	Sales report

manufacturers only like to use one grade of material on a particular machine to avoid cross-contamination from tools.[2]

Do not assume "6.2 – Animal products" is a simple item (Table 5.6). You really need to ensure that you have no animal products in your manufacturing chain – they can creep in anywhere, especially with plastics. It may not be universally known, but the lubricants used in plastic production, machining, and extrusion can be of animal origin. You *must* get confirmation from your suppliers that this is not the case.

5.3.1.7 Packaging and Transportation

This is simple but often overlooked. Too many times I have heard "If only it were 12 mm shorter it would have fit in our standard box." There is so much to the final packaging of a device that it must not be ignored (Table 5.7). What packaging is required: sterile or nonsterile? What labeling is required? What is to go into the package – full instructions or single instructions for use sheets? What size box should it go into?

Does the packaging have to be specially designed to meet the rigors of transportation? You will be amazed at the vibration a device has to withstand just in the boot (trunk) of a

---

[2] When you machine a material some of the machined material resides on the tool itself. Hence if you machine titanium one morning and then stainless steel in the afternoon you may impart some titanium on the workpiece – hence contamination is possible.

Table 5.6: An Example of the Manufacturing Section in a PDS

Section 6: Manufacturing		
Number	Comment	Source
6.1	Cleaning: Device needs cleaning post-manufacture.	
6.2	Animal products: Manufacturing to be free from any animal by-products in the manufacturing process.	
6.3	ISO 9001: Company rules limit suppliers to those who have ISO 9001 or ISO 13485 certification.	Company rules
6.4	Stainless Steel – 316LVM: Company rules state that all SS should be 316LVM.	Company rules

Table 5.7: An Example of the Packaging and Transportation Section in a PDS

Section 7: Packaging and Transportation		
Number	Comment	Source
7.1	Drop test: When packed the device should withstand a drop from a height of 1 m.	Standards review
7.2	Instructions for use: Each box to contain 1 IFU.	MDD – FDA
7.3	Shelf life: The shelf life for the packaging is to be 3 years.	Focus group report
7.4	Packaging dimensions: To fit existing box, a footprint of 100×200×50 mm.	Packaging office
7.5	Labeling: Keep package upright, label required.	Packaging office
7.6	Assembly in situ: Assembly on delivery to be minimized.	Focus group
7.7	Vibration: Packaging to insulate device from vibrations induced in road transport.	Standards review

car. Do not underestimate the trauma a device goes through moving from A to B. You may well have to specify tests to ensure your packaging is correct (these are called *accelerated life tests*).

Also, do not forget that the packaging has to fit into a vehicle of some kind. It is so easy to forget this simple basic concept and then find out that your design doesn't fit into your biggest van or, even worse, will not fit into a standard shipping container. Talk to your distribution arm, talk to your shippers…it really is that easy.

5.3.1.8 Environmental

The green agenda is now written in law in virtually every state. We all have a role to play in the recycling and disposal of waste (Table 5.8). At present the healthcare system is not very green, but that does not mean that we can't be. So long as we meet all the requirements laid down in Section 5.3.1.2 we can minimize waste. I am sure this section is going to grow as people start measuring waste by sector and identify healthcare as a significantly nongreen industry. Hence I am sure medical devices manufacturers will soon be determining carbon footprints, putting recycling labels on outer boxes, and using recycled cardboard.

However, don't forget that some devices actually rely on the use of very harmful substances that emit noxious fumes or electromagnetic radiation. In these circumstances we are not immune to environmental regulations and our design will need to ensure that they have been met.

Table 5.8: An Example of the Environmental Section in a PDS

Section 8: Environmental		
Number	**Comment**	**Source**
8.1	External packaging: All external packaging to be recyclable.	Focus group report
8.2	Internal packaging: No recycled packing materials may be in contact with the device.	Focus group report
8.3	Disposal: Disposal of the device limited to "sterile sharps" restrictions.	Focus group report
8.4	Servicing: By-products of servicing to be disposed of meeting the requirements of relevant standards.	Standards review

5.3.1.9 Summary

It is important that you appreciate fully the value of a good PDS. While the above eight sections have tried to demonstrate what a PDS contains, only the actual development of a full PDS demonstrates the full beauty of its structure. We will produce one later in this chapter as a case study; but before we do we need to look at where the information comes from. The important thing to remember is that you should spend as much time as you think necessary to produce a viable PDS, but also remember it is a living document and it can be modified as your design develops.

5.4 Finding, Extracting, and Analyzing the Content

This section will look at the ways you can determine the content of the PDS. It is not split into the PDS sections; each of the following items could supply information for any of them. Also note it is applicable to any information sourcing throughout the whole design process.

Figure 5.4 illustrates a model that I call the *data cloud* and how it interacts with the procedure for the development of a PDS. The concept of a data cloud is a good one: the sources are numerous, they float around in the space around you, they are amorphous, and they are often just out of reach. Your role, as the designer, is to attack all of these sources and determine which aspects of each of the sources are relevant to your proposed design. Nearly all of them will have some input, but some will be more direct than others. In the following sections I will attempt to show you how to make sure that each of the individual voices in the cloud influences your PDS.

5.4.1 Focus Groups

Focus groups are a collection of end-users or "stakeholders"[3] in a meeting place of suitable standing. It is quite normal for your sales force to have identified end-users who are sympathetic, free speakers and who do not bear grudges against specific disciplines. It is also quite normal for you to have your own list. The basic premise of a focus group is to bring a group of people together to discuss the issues around a topic of interest with a view to determining the solution to a particular issue (or issues). You will almost certainly need to consider the production of a non-disclosure agreement[4] to ensure that confidentiality is ensured and maintained.

The basic goal of the focus group is to start the discussion concerning customer requirements. Because the focus group is essentially a group of "friends," they can be trusted to give you

---

[3] Stakeholders are people directly concerned with your device who may not be end-users but who have defined links to the discipline, use, purchase and, specification. They are not holders of shares in the company.

[4] A non-disclosure agreement (NDA) is a legally binding document that is signed by all parties to ensure that members of the focus group do not exploit what they have heard, nor share it with anyone else. If you do not have an NDA you need one now!

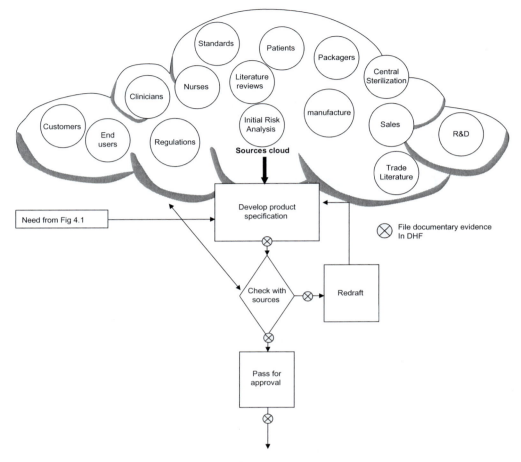

Figure 5.4
The data cloud and its interaction with the PDS.

unbiased opinions concerning the requirements for a device. However, they may not be totally au fait with the subject or discipline hence the comments they make are not written in stone and should be taken at face value until confirmed. It is sometimes worth having one antagonistic focus group member who plays "devil's advocate"[5] to ensure that both sides of an argument are explored.

There are many ways that focus groups are run and there is no model of good practice. However it is a truism to say that good food and libation normally sets tongues wagging

---

[5] A "devil's advocate" is a person who is given the specific task of questioning a statement irrespective of their belief in order to give fair hearing to both sides of an argument.

hence your selection of a venue will make a great difference to the outcome. You should, however, be wary of bribery and undue influence laws. In no way should your invitation to the focus group be formatted in such a way that it looks like you are trying to influence a clinician to purchase your products; this is crossing the line and can open you to a bribery conviction. So stay clear of fully paid holidays in Barbados for the whole family. You should not be questioned for organizing a focus group where reasonable expenses have been paid for.

Although a focus group sounds unstructured it should not be; it must be well planned and well executed. For the group's first meeting an "icebreaker" should be the first item on the agenda; this may be drinks and dinner the night before or it may be a structured activity. Your agenda is explicit: stay focused, stay alert, and make sure you keep clear notes.

The focus group's potential activities are numerous but several good examples follow:

> Post-it SWOT: Conduct a SWOT (Strengths Weakness Opportunities Threats) analysis by allocating one, or more, wall(s) to each and giving the participants Post-it Notes. They write the S, W, O, or T on the note and stick it to the appropriate wall.
> Round table: A simple round-table discussion of a topic. The topic/question must be well written and the discussion must be well chaired.
> Magic ball: A question is set, such as "What color should it be?" A member of the group can only speak when they have the magic ball in their hand (obtained by holding their hand up). This is a good idea when there are people speaking over others.
> Free table: Simple discussion over dinner and into the evening. Guests are interspersed with "spies" who listen, note, and nudge the conversations along.

Make sure you pick the right method for your focus group. An activity that works with one clinical group may not work with others. Make sure that you also do a cost–benefit analysis to ensure that you are not paying too much!

Do not fall into the trap of having a group full of "yes-men." You need to have critical evaluation so while you do not want to have a room full of ogres the odd negative comment is not worthless and should not be feared. All comments add to your specification; all voices are relevant.

Always remember that some of the end-users and stakeholders will know more about the relevant standards and industry norms than you do. Focus groups are a great way to find out about relevant regulatory items without giving away the fact that you may not know them all.

After each focus group make sure you perform a full debrief with those who have helped you to run the session. You should have allocated someone to take notes throughout – shorthand is a very valuable skill! Everyone's collective recollections, thoughts, and notes should be collated, analyzed, and filtered into a single brief report highlighting the issues that the PDS needs to take into account.

5.4.2 Regulatory Bodies

There can be little doubt that these are an unmissable first port of call, either directly before or directly after talking to the customer. Each of the main regulatory bodies has guidance documents, copies of relevant statutes, and even people at the end of a telephone. All of which is a valuable library of information.

Taking the FDA website as an example (www.fda.gov) one of the most useful items is the 510(k) search. This single item enables you to identify any previously approved devices that may be similar to yours, and as a consequence reveals valuable information about classification, etc. The UK's MHRA website (www.mhra.gov.uk) hosts all of the guidelines, documents, and links to documents that enable you to start building your specification in relation to CE marked devices.

In relation to standards the FDA hosts a database of standards where consensus has been reached. In other words, designing to a cited standard means that it is recognized by the FDA. Just opening and examining the database gives you a valuable starting point for your standards review – once again a valuable resource.

Many people overlook the recall, vigilance, and notifications databases. These give fantastic insights into failure modes…allowing you to learn by the mistakes of others.

5.4.3 Immersion

One of the best ways to find out what a design needs is to immerse yourself in the area. An actor would call this *method acting*. If the device is to be used in an OR theater then try and get into one; if it is to be used by nursing staff on a ward, go to one and observe. Irrespective of the complexity of the design, immersion into the environment is second to none. Recently one of my students was working on the design of a wheelchair attachment; I made him spend a day in a wheelchair to understand the environment from the perspective of the end-user. That day was worth its weight in gold!

If you can get into the actual environment you will learn so much about the requirements. You will identify the questions to ask. Another reason is to learn "the language" of the environment. Often terms used in the clinical context can be different to those used in others. For example the word "distraction" clinically means moving two items apart, not something that disrupts your concentration. You have to immerse yourself in the environment to understand the language just as you would to learn a foreign language.

Sometimes immersion is hard to achieve – for example you may not be able to witness an operation directly. However training videos are always available as are clinical textbooks. These do not give the full picture but will enable you to think of the right questions in a focus group.

This is a good basis for your investigation report. Using your experiences and recollections of your period of immersion and tying this up with your own expertise and some further personal research develops entries for the PDS that would not otherwise exist. You should produce a brief investigation report similar to that of the focus group.

5.4.4 Libraries

None of us are too far away from a library. In the modern Internet age libraries are available online. All colleges and universities have well-stocked libraries and excellent access to journals and standards. Most will be more than willing to help and it does not hurt to ask. Equally, county, state, and national libraries are generally free and the staff is generally helpful. You should also try to develop your own library by collecting and collating standards, papers, and textbooks to which you refer on a regular basis. Let us examine the sort of items you should refer to.

5.4.4.1 Standards

In general you should refer to ISO (International Standards Organization) documents as your first point of call. National bodies such as ASTM in the USA and BSI in the UK both have their own standards and online search engines that enable you to identify relevant standards from keyword searches. There is little doubt that you will have to purchase a copy of the main standards (e.g., ISO 13485) and those you refer to on a regular basis. However they are expensive and to purchase them on a whim is not viable, hence the use of libraries to review up-to-date copies is advisable.

The format of a standard is not unique but there are certain things you should know to enable you to make full use of them. Normally the title of a standard is very explicit and makes complete sense – however the title does not cover everything. Many standards are multi-numbered (especially ISO); their numbers refer to the standard they equate to in a given state. Don't forget to use the FDA equivalence database to check.

Quite often a standard is split into "parts" to make it more legible. The parts may not all be of the same year of publication, which can be confusing. Here the *scope* becomes important; the scope states what the standard actually covers. Associated with the scope is a list of referenced/associated standards – these help your research to expand organically. It is important to know that standards are still current; they are superseded and withdrawn on a regular basis so always check before use. All of this before we have even read the standard itself! When citing a standard in your PDS give a full reference: title, standard number, section, and page number. All this helps the people who follow to find the relevant document quickly. As with previous sources, produce a brief report to help with the compilation of the PDS.

5.4.4.2 Journals and Learned Publications

There is little doubt that your design will be influenced by the current state-of-the-art. Learned publications are one of the sources of this information. These are scientific journal papers (often 9– 0 pages in length) housed in specialist magazines called *journals*. The papers are *peer reviewed* (this means someone else has checked the content and agrees it is correct) and they should be number one on your list, well above web-based documents. Your particular device will fit within a clinical discipline so identify the discipline and the journal(s) that go with it (the focus group will help here). Now your links with the university libraries will bring forth fruit as they will most likely have copies of the journals. If not you can use one of the many scientific search engines to find the paper and procure it. Some valuable search engines that you have access to are:

Google Scholar: http://scholar.google.co.uk/
PubMed: http://www.ncbi.nlm.nih.gov/pubmed
Science Direct: http://www.sciencedirect.com/

Some papers are free to access, some you will have to buy; all have abstracts (or summaries) that are free to examine. Eventually, you will isolate the journals you are always interested in and it is possible to receive emails of the contents when published.

Reviewing the scientific literature is the basis of your literature review. The papers you deem to be important to the PDS should be kept intact and summarized in the review. This review needs to be a brief report that the PDS can refer back to; again, to make life for those who follow easier.

I keep refering to "those who follow." What do I mean by this? It is important to remember two things. Firstly, you may not be doing the actual design, you may only be producing the PDS, hence the designer that follows should not need to keep coming back to you to ask questions about sources of information. Secondly, and this is rather bleak, you could have an accident and expire. In this case there is no one to ask; hence a fully documented PDS is absolutely essential. You may not expire, you may just leave the company; to the designer following on it's the same thing.

Another important aspect of journals is the concept of citing. Every paper will have a list of references at the back; these are publications the authors think are worthy to refer to. If they are worthy of being read by the authors are they not worthy of you too? This is called a *citation review*. Effectively you find the most modern paper you think is important and work back in time. Soon you begin to find the common books, papers, and publications that people refer to; these are often the best sources to find out about core issues. Modern web resources (such as Google Scholar) do this for you.

5.4.4.3 Books
Libraries are the natural home of books: textbooks, reference books, encyclopedias, and historical texts. Although we are in the Internet age we still need to refer to bona fide sources. While the text on the web remains unregulated we cannot rely on content. Books are still a

mainstay. Luckily, there are now electronic libraries one can subscribe to that make access easier, but I can guarantee that for the near future you will not escape the textbook no matter how hard you try. As with other sources, give the full reference using the Harvard standard and the page number.

5.4.4.4 Librarians

Do not underestimate the knowledge of the librarian. Often, and in universities in particular, they are highly knowledgeable in their subject discipline. They can point you in the right direction and are normally happy to do so; just ask.

5.4.5 Technical Literature

We all receive magazines in the mail. Some are rubbish, some are valuable, and some are good. They are all valuable sources of information. We all go to exhibitions where companies give away materials such as catalogs and fliers. We are all able to go to the relevant clinical exhibition to collect trade material related to your design space. Remember Newton said "I stand on the shoulders of giants."

5.4.5.1 General Trade Magazines

These arrive unsolicited through the mail and contain trade articles and advertisements. It is the trade articles we are interested in as they may contain a nugget of information related to your design. While you cannot keep everything, you should retain what you think is useful. In other words, think about a file of recipes cut out from a domestic magazine. It looks useful, and one day it will be used.

Do not be afraid to receive trade magazines.

5.4.5.2 Catalogs, Fliers, and Trade Literature

Simply build your own library. Keep everything up to date and logically filed. As with the cuttings from trade magazines you never know when one will come in useful.

5.4.6 The Internet

Beware of this source! Because it is unregulated you can download information that looks bona fide but which has actually been produced by some illiterate, grotesque Hogarthian deep in a hole somewhere in the middle of nowhere.

Only use sources that you know are genuine (such as the journals and online libraries described earlier). At the end of the day use other Internet sources with a keen sense of disbelief.

5.4.7 Conferences and Symposia

Let us end on a high note. We all need to get out of the office sometime; so why not go to a conference. There is a plethora of meetings, conferences, short courses, and symposia in your subject area. They can range from general medical devices exhibitions (such as MEDICA in Germany) to discipline subject meetings (such as the Foot and Ankle meeting). All are valid; but you have to be discerning.

If you want to meet the current thinkers in your design space go to the relevant meeting. Some will be highly clinical (those organized by medical bodies); some will be highly scientific (organized by scientific/engineering bodies); and some will be aimed at the trade (organized by trade organizations). Some will be attended by 100,000 people, some only 10. The main thing is to plan what you are going for: What are you trying to find out about? What is it you want to know?

If you go to a conference, download the program of talks and pick the ones you really want to go to. Go, listen, and take notes. You may even have found a member for a focus group! Make sure you get a copy of the conference proceedings. At conferences some people put up posters, sometimes photographs are allowed, sometimes not…you need to check beforehand.

Conferences are an excellent venue for a focus group; the people you want will probably be going anyway. If you need to invite somebody, what is better than to invite them to participate in a conference? Equally, annual trade meetings are also good venues for focus groups, for the same reasons.

5.4.8 Others

While I have listed many sources they are by no means a definitive list. You are open to use all of your intelligence, whiles, and guiles to determine the information you want. Just remember three main rules:

> It has to be legal.
> It has to be reputable.
> It has to be documented.

CASE STUDY 5.3

In this section we will look at a case study to enable you to see the basics of a product design specification. I have tried to make the PDS complete but no doubt you will find some missing items, or you may disagree with some of the terms. Do not worry, it is not meant to be a model answer!

In this case study we shall develop a PDS for a simple drill bit for drilling into bone. (This example has been selected as it is easy to imagine the outcome.)

After holding discussions with the surgical team and OR theater staff the following points were critical: it must be 4.8 mm diameter; it must have a stab point; it must be reusable; the flutes should be a minimum of 50 mm long; and the overall length should be 200 mm.

Table 5.9: Example PDS for a Bone Drill Bit

Medical DeviceCo Inc.
Product Design Specification   Originator Date
Project Number/Part Number   Drill 200048050S   Version: 1.0
Product Title: 200 mm×4.8 stab point drill 50 mm flutes
Summary
This specification is for a reusable drill bit for producing 4.8 mm dia. holes in human bone.
It is estimated that this device is a transient, invasive device but is one that is *reusable*; hence it is a *reusable surgical instrument class I*.

1 Customer:		
1.1	Hole diameter 4.8 mm	Initial focus group
1.2	Overall length 200 mm (nominal)	Initial focus group
1.3	Flute length 50 mm (nominal)	Initial focus group
1.4	Device to be reusable	Initial focus group

2 Regulatory and Statutory:		
2.0	Device to meet essential and general requirements of a medical device	
2.1	Initial estimate is that this is a reusable surgical instrument and hence CE/FDA Class I (510(k) exempt)	93/42/EC Annex IX rule 6 CFR 21 Reg. No 888.4540
2.2	Drill material to be selected from those approved in standards	ISO 7153 ASTM F899-09
2.3	Labeling to show it is supplied nonsterile	93/42/EC
2.4	In EU labeling to comply with standard	93/42/EC BS ISO 15223-2:2010
2.5	In U.S. labeling to comply with regulations	CFR 21
2.6	IFU to be supplied with each drill and to include cleaning and sterilization instructions	93/42/EC CFR 21
2.7	Declaration of conformity required	93/42/EC

3 Technical:		
3.1	Flute helix 14° (nominal)	Trade review
3.2	Material to withstand high alkalinity (pH 13–14) in washers	Cleaning and sterilization review
3.3	Material to withstand +130°C in steam sterilizers	Cleaning and sterilization review

4 Performance:		
4.1	Drill time in bone to be no worse than existing 4.8 mm drills	
4.2	Drill to perform consistently for 25 individual uses	
4.3	Breaking torque to be no worse than existing 4.8 drills	
4.4	Bending strength to be no worse than existing 4.8 drills	
4.5	"Point" to locate hole securely to within ±1 mm	

5 Sales:		
5.1	Sales cost c £40 (gross margin 60%)	Sales report
5.2	Estimate 100 sold per month	Sales report
5.3	Should fit in std Jacob's Chuck	Sales report/follow-up focus group
5.4	Packaging to be minimal	Sales report
5.5	Would be good to have a mark every 5 mm between 60–90 mm to estimate drill depth (UPS against other drills)	Sales report
5.6	Can the flutes be gold in color to match market leader?	Sales report

6 Manufacturing:		
6.1	Invasive hence restricted to ISO 13485 subcontractors	Company policy
6.2	Device to be supplied clean in batches	Standards review
6.3	Sharp edges to be protected	
6.4	Finish to be to standard	Standards review ASTM F86-04 ISO 9714-1:1991 BS 3531-5.5:1990 BS 7254-2:1990
6.5	No animal products to be used in manufacturing	
6.6	Device to be laser marked with CE mark, dia., company logo, part number, and lot number.	Standards review 93/42/EC ASTM F86-04 ASTM F983-86
6.7	Materials restricted by standards	ISO 7153 ASTM F899-09

7 Packaging and Transportation:		
7.1	Supplied to end-user as single items	Initial focus group
7.2	Packaging to protect stab point	
7.3	Packaging to protect sharp cutting edges	
7.4	Standard tube 200×10 mm	
7.5	Label to state nonsterile	93/42/EC CFR 21
7.6	Label to state manufacturer's name, date of packaging, CE mark, lot number	93/42/EC CFR 21 BS ISO 15223-2:2010

8 Environmental:		
8.1	Packaging to be recyclable (if possible)	
8.2	Invasive device, disposed as clinical sharp	

Approved/Not Approved
Signed
Date

PDS.doc   version 1.0   Approved by:
                                              Date: 17.5.2010

Case Study 5.4

Using the information gathered for Case Study 5.3, repeat the exercise but ignore the fact that they have asked for a drill bit – imagine that they have requested an item to produce a hole of 4.8 mm dia. but do not stipulate a method. How does this change the PDS?

Notice that this case study removes the "sacred cow" of a drill bit. There are many other ways to produce a hole…and this PDS does not limit the solution.

Table 5.10: Example PDS for a Device to Produce a 4.8 mm dia. Hole

MedicalDeviceCo Inc.
Product Design Specification   Originator PJO Date 21/8/11
Project Number/Part Number   04801   Version: 1.0
Product Title: Device to produce a 4.8 mm hole
Summary
This specification is for a reusable device for producing 4.8 mm dia. holes in human bone.
It is estimated that this device is a transient, invasive device but is one that is *reusable*; hence it is a *reusable surgical instrument class I*.

1. Customer:		
1.1	Hole diameter 4.8 mm	Initial focus group
1.2	Overall length 200 mm (nominal)	Initial focus group
1.3	Device to be reusable	Initial focus group

2 Regulatory and Statutory:		
2.0	Device to meet essential and general requirements of a medical device	
2.1	Initial estimate is that this is a reusable surgical instrument and hence CE/FDA Class I (510(k) exempt)	93/42/EC Annex IX rule 6 CFR 21 Reg. No 888.4540
2.2	Material to be selected from those approved in standards	ISO 7153 ASTM F899-09
2.3	Labeling to show it is supplied nonsterile	93/42/EC
2.4	In EU labeling to comply with standard	93/42/EC BS ISO 15223-2:2010
2.5	In U.S. labeling to comply with regulations	CFR 21
2.6	IFU to be supplied with each item and to include cleaning and sterilization instructions	93/42/EC CFR 21
2.7	Declaration of conformity required	93/42/EC

3 Technical:		
3.1	Material to withstand high alkalinity (pH 13–14) in washers	Cleaning and sterilization review
3.2	Material to withstand +130°C in steam sterilizers	Cleaning and sterilization review

4 Performance:		
4.1	Hole production time in bone to be no worse than existing 4.8 mm drills	
4.2	Hole production to perform consistently for 25 individual uses	
4.3	Breaking torque to be no worse than existing 4.8 drills	
4.4	Bending strength to be no worse than existing 4.8 drills	
4.5	"Point" to locate hole securely to within ±1 mm	

	5 Sales:		
5.1	Sales cost c £40 (gross margin 60%)	Sales report	
5.2	Estimate 100 sold per month	Sales report	
5.3	Should fit in std Jacob's Chuck	Sales report/follow-up focus group	
5.4	Packaging to be minimal	Sales report	
5.5	Would be good to have a mark every 5 mm between 60–90 mm to estimate hole depth (UPS against other drills)	Sales report	
5.6	Can it be gold in color to match market leader?	Sales report	

	6 Manufacturing:		
6.1	Invasive hence restricted to ISO 13485 subcontractors	Company policy	
6.2	Device to be supplied clean in batches	Standards review	
6.3	Sharp edges to be protected		
6.4	Finish to be to standard	Standards review ASTM F86-04 ISO 9714-1:1991 BS 3531-5.5:1990 BS 7254-2:1990	
6.5	No animal products to be used in manufacturing		
6.6	Device to be laser marked with CE mark, dia., company logo, part number, and lot number.	Standards review 93/42/EC ASTM F86-04 ASTM F983-86	
6.7	Materials restricted by standards	ISO 7153 ASTM F899-09	

	7 Packaging and Transportation:		
7.1	Supplied to end-user as single items	Initial focus group	
7.2	Packaging to protect stab point		
7.3	Packaging to protect sharp cutting edges		
7.4	Label to state nonsterile	93/42/EC CFR21	
7.5	Label to state manufacturer's name, date of packaging, CE mark, lot number	93/42/EC CFR 21 BS ISO 15223-2:2010	
7.6	Standard package sizes given in packaging register	Stock package register	

15 Environmental:		
8.1	Packaging to be recyclable (if possible)	
8.2	Invasive device, disposed as clinical sharp	

Approved/Not Approved Signed Date

PDS.doc   version 1.0   Approved by:
                                                Date: 17.5.2010

You should see that little has changed. All that has been removed is the word "drill" where it is not required. Now your design is free to select any hole production method from punches, broaches, lasers, borers, water jet – anything you can think of, even drill bits. Your design process will pick the best and most appropriate solution.

Hopefully, from the above, you will see that the wording of the specification need not be lengthy. The statements should be brief, concise but informative. They should not leave things open to conjecture and, wherever possible, they should point the reader to the source of further information. Notice also that the specification does not provide solutions, it only provides the boundaries for the design space. I know this was a drill bit, and hence the image is in your mind, but this was intentional so that you can imagine the PDS – if I had picked something too abstract you would not have been able to relate to the PDS.

The main thing to recognize is that the next person in the trail could lift this document and produce the design without asking any further questions. They will need to refer to the documents stated in the sources, but that is good practice. This is not a drafting exercise, there are still things to decide and some design choices to be made.

The other thing to note is that a good PDS will satisfy all auditors that you have clearly investigated your inputs and hence have met the requirements discussed in Chapter 4.

5.5 Summary

In this chapter we examined the process to develop a statement of need and a full product design specification (PDS). We saw that the PDS is influenced by sources in the data cloud and that you, as the designer, have to use every tool at your disposal to gain access to the information.

We looked at the sources in detail and we also looked at methods to conduct efficient research. It was concluded that for each section of the PDS a brief report describing where the information came from should be written.

We also saw that we need to document the process fully. This was suggested for two main reasons: the first being that you may not be the final designer; the second being that you may leave the project either intentionally or through accident. In both situations future designers need the information stored in you head.

References

Bicheno, J., & Catherwood, P. (2005). *Six sigma and the quality toolbox*. Buckingham: PICSIE Books.

British Standards (2001). *Product specifications. Guide to identifying criteria for a product specification and to declaring product conformity*, BS 7373-2:2001.

Enderle, J., & Bronzino, J. (2011). *An introduction to biomedical engineering* (3rd ed.). Academic Press.

European Community (1993). *Medical Devices Directive*. 93/42/EC.

FDA (1997). *Design control guidance for medical device manufacturers*. FDA.

FDA (2010). 21 CFR Subchapter H, Part 860.

ISO (2003). BS EN ISO 13485:2003 *Medical devices – Quality management systems – Requirements for regulatory purposes.*

ISO (2008). ISO 9001:2008 *Quality management systems.*

Hurst, K. (1999). *Engineering design principles*. London: Arnold Publishers.

Pugh, S. (1990). *Total design: Integrated methods for successful product engineering*. Prentice Hall.

Ulrich, K. T., & Eppinger, S. D. (2003). *Product design and development*. McGraw Hill.

Webster, J. (2009). *Medical instrumentation: Application and design*. J Wiley and Sons Ltd.

## CHAPTER 6

Generating Ideas and Concepts

6.1 Introduction

In the divergent–convergent model, described previously, we saw that the generation of ideas and concepts was of paramount importance. The specification developed in the previous chapter should result in numerous ideas that are able (or unable) to meet its requirements. It is not possible to overstate the importance of being open to the generation of ideas. It is the single weapon against the "sacred cow."

In this chapter we shall be looking at tools and methods that enable ideas and concepts to be generated, liberally. Why liberally? One important aspect of design is to select the best solution that meets a need...how are we able to select this from only one idea? We need as many as possible. Consider golfers, do they go onto a golf course with only one club? No, they do not – they have a whole bag full. Each one is a potential club to use but only one is ideal; their first job is to pick that "one" club (remember the "one thing" quote). However the analogy breaks down because they are able to fill a bag of clubs from a shop. We do not have a shop of ideas we can walk into; however we can do our best to create one.

Hence the aim of this chapter is to give you the tools that enable you to build that "shop of ideas" so you are able to pluck potential ideas and concepts off the shelves, at will. Some of the tools you can use on your own, for some you will need to be in a group, and some can be either. I have attached symbols next to the section title to make identification easier:

On your own: This symbol means this activity can be performed on your own.

In a group: This symbol means this activity is, or can be, performed in a group.

6.2 The "Engineer's Notebook"

If any of you have studied art you will have been told to carry a sketch pad with you at all times. The same concept applies to designers, but we have a "notebook." You will be amazed how many ideas come into your head at the weirdest times and in the most obscure locations.

Medical Device Design.
DOI: http://dx.doi.org/10.1016/B978-0-12-391942-7.00006-4

Songwriters, for example, often get ideas in their sleep and when they wake have to get them down as soon as possible. Hence they have a notebook by the bed. Do you think Van Gogh or Constable were devoid of their sketch pad? You should think of yourself at their level; you are the Van Gogh of medical device design, hence you will use your "engineer's notepad" and keep it in hand at all times.

I would like to share an anecdote as an example. I was at an annual meeting recently with a colleague of mine, when we met someone we'd met the year before. I was amazed when she enquired after his wife and children by name. When I asked how on earth she remembered their names, she let slip the trick of all professional "net-workers." After a party, meeting, or soirée they write down the names of any interesting people (with notes) in a notebook. This way, if they meet again, it looks like they remember every detail about them. In this case she was expecting to meet him at the meeting so looked back in her notebook. Apparently this is commonplace in PR, etc. So why not use this trick in design? If you have an idea note it down, but you need something to note it in.

As you meet different people you will pick up little tips, little rules of thumb. It is hard to remember them all, so have a little notebook in your pocket and write them down. Sometimes you will be relaxing in a bar and something, out of the corner of your eye, gives you an idea – out pops the notebook. If you don't get the idea down as soon as possible it will be lost and you will never recall it. You will even get ideas in the shower, in the bath, and on the toilet. You will soon learn how invaluable your little notebook comes to be. Nowadays we have lovely electronic toys such as smartphones, laptops, and iPads, all of which can be your "notebook." Personally, I still prefer my little pocketbook and pencil.

6.3 Creative Space

There is little doubt that the best creative thoughts are created when the environment is applicable. Note I said applicable, not best, not ideal, not stupendous, just applicable. In this section we shall try to determine what "applicable" means. I would love to point you to some research that helps you create a space but we will soon start to move into feng shui. However, there are lessons to be learned from feng shui as it is all about designing your room to create energy…that is what this section is all about.

6.3.1 The White Room

If you do a Google search for "white room" you will have pages upon pages of design agencies called *white room design* or *white room agents*. This is because the white room concept is drummed into every design student while at university. A white room signifies the blank page. It is supposed to convey the concept of a room in which anything is possible; you are starting with a "clean sheet." Some design agencies have the space, and the money, to build an actual white room in their office block – not all of us are so lucky. However we can use the concepts of the white room (or white room rules) everywhere:

- People "entering the white room" are equal – there is no hierarchy and everyone's point of view, idea, or comment is as valuable as another's irrespective of their actual position in the company or in life.
- All ideas are valid – no idea is destroyed, pooh-poohed, or thrown out. All are kept and analyzed later.
- Open discussion is promoted – free speech is an absolute must and no one is allowed to speak over any one else.
- The "white room" is a space free of "clutter" – no posters, no distractions.

If you apply these basic rules you can have a white room in your own house, in your attic, even in the local bar. Irrespective of whether you use a white room or not, the four bullet points above should be adhered to in all group-based activities. In *Cracking Creativity*, Michalko (2001) points to research that states that the great minds of Einstein, Bohr, Heisenberg, and Pauli were able (even with their accumulated egos) to share and discuss ideas openly, freely, and informally because "their discussions were open, free, and spontaneous."

So in essence the white room is a space in which to develop and capture concepts and ideas from a group of people. Most authors suggest that the groups should be about five or six people, but I have done sessions with as little as three and as many as 10; 50 would be silly!

6.3.2 Personal Space

This is where you may fall out with your boss (if you have one). I always find it weird that in many companies the CEO will have designed their own office, picked their own furniture, selected their own pictures, and even decorated the office to their taste, while the employees have a predetermined sterile layout. If you are to be creative you need to feel comfortable, that is why lots of ideas come, for example, when you relax in a bath. Some of you may be quite happy sitting at a desk, but this does not get the creative juices flowing. Everyone needs some creative space. My office at university, for example, has an old leather recliner chair, a real stereo hi-fi with proper amplifier and speakers, and low lighting. When I need to think, the lights go low, I sit in the recliner, and I put on some nice music and relax – you will be amazed how this simple routine works. On my wall I have a small poster of Audrey Hepburn, and when I get stuck I just turn around and look at her and ask, "Come on Audrey what would you do?": silly, I know, but it works for me! Also, I get told off for being untidy. That's because when I am being creative I keep all of my ideas in heaps around me – when the project is finished the papers are filed, the office tidied, all ready for the next one. For some that is annoying and they say "How can you work like that?" A colleague of mine never ever had any pieces of paper on show, at any time: I used to ask him "How can you work like that?" Personal space is just that, personal. We are all different; space cannot be imposed (bosses, CEOs, and supervisors take note). Many modern companies (Google for example) have revolutionary workspaces.

Grossman (1988) says, "being creative boils down to having fun." He talks about two kinds of motivation: *intrinsic* and *extrinsic*. Creative space can stimulate. Apparently increasing extrinsic motivation with an attractive and stimulating place offers only short-term benefits. On the other hand the intrinsic motivation your space provides yields longer-term results. Its physical properties should be flexible; it should be adaptable so that the designers see it as a product of their own making. More importantly the space should reflect and support the needs of its users, and its use.

> *"The more flexible and adaptable your design, the longer you will see both intrinsic and extrinsic motivational rewards."* (Lloyd, 2011)

This is nothing new; at the turn of the twentieth century Henri Fayol was looking at workers in an office space with a view to motivating productivity. He thought, and I paraphrase, that including some "nice things" in the workplace would produce benefits, so they introduced some and productivity and general "happiness" increased – but tailed off. So they introduced some more "nice things"; productivity increased – and tailed off. In the end they took all of the nice things away and went back to the same space as before – productivity increased again! He found that the people were motivated by the "changes." People like to feel like they belong, like they have ownership…so why not make them feel so?

The lesson here is that your creative space needs to be *your* creative space. You need to feel comfortable; it needs to be stimulating and amenable to change. Utmost and foremost it needs to work – for you.

6.4 Generating Concepts/Ideas

This next section is concerned with tools that you can use to generate ideas and concepts. They are not a complete list and you should refer to the library as much as possible. Three texts that are useful starting points are by Hurst (1999), Dym and Little (2000), and Ulrich and Eppinger (2003).

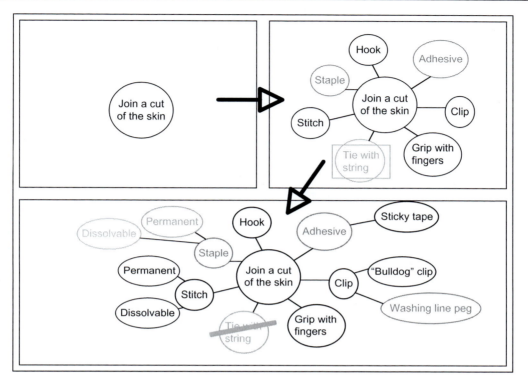

Figure 6.1
An example of *radial thinking* used to determine ways to join together a simple cut.

6.4.1 Radial Thinking

This is my own little extension to "mind mapping." The idea here is to develop ideas within a structure that has no structure. Sounds crazy I know, but that is what it is. The single thought or problem you need to solve is written in a small circle in the middle of a very large sheet of paper (be prepared to start with a sheet of A1 and to stick sheets of paper together). Everyone is given a different color pen. First rule: you can add to the diagram at will. Second rule: no one is allowed to delete an entry – only the person who wrote it down can strike through it (with a single line and adding a note why it is being deleted) after persuasion from the other group members (note persuasion by discussion not bullying). A line is drawn between the circles. This will build outwards in layers building a picture of potential ideas and solutions.

Figure 6.1 illustrates the progression of the method. Note that you are not restricted to using "correct terminology." Many people in your group may not have your background or that of your clinical staff; this should not be allowed to inhibit the thought process. Just because someone doesn't know the right word does not make it a bad idea. Be very aware of intellectual snobbery.

6.4.2 Inversion (or Word Association)

This is a very simple technique used to promote lateral thinking. The process is simple; write down a word or phrase associated with the concept, e.g., white, then write down the opposite (or its antithesis), i.e., black.

If I may I will give you an example from the diesel engine industry. All cars, lorries, and trucks have a fuel line to the engine. One particular fuel line was vibrating in use at a particular frequency F and failing (Figure 6.2). The obvious answer was to add more clips to stiffen it. However, inverting the solution meant taking away a clip to make it more flexible. Both solutions worked, but the removal of a clip was the cheaper solution as they were making thousands of engines per year.

Hence inversion simply examines the opposite. If someone says it should be "stiff," also write down "flexible." If someone says it should be "fast," also put down "slow." Some examples are shown in Table 6.1.

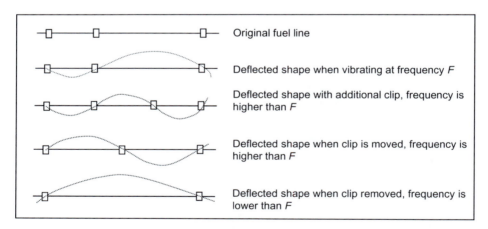

Figure 6.2
Example of the inversion technique used to a solve vibration problem with a fuel line.

Table 6.1: Example of Inversion Technique

Specification Inversion	Slow Fast	Stiff Flexible	Dark Light	Smooth Rough	Shiny Dull

This technique does not directly lead to a solution, but enables you to get rid of "sacred cows" and start to think laterally. The obvious extension to this is to also use word association as sometimes a different word sparks an idea, for example "stiff – rigid."

6.4.3 Analogue

Analogue (noun) – a thing, idea, or institution that is similar to or has the same function as another.

The idea is to think of an analogy that you can attribute to the problem you are trying to address. Analogues are commonplace in theoretical modeling of systems that are hard to understand – the first port of call is to find something that is similar (its analogue). The analogues are nearly always from the physical world and, hence, drawn from experience. If, for example, you were trying to design a frame to aid ambulation for persons with disabilities you may look at other things than humans walking – any biped would do, e.g., penguins, cranes, gorillas, etc. (Figure 6.3). This is a very powerful idea generation method as it makes you (excuse the jargon) think outside the box.

As a further example of analogues there is a new engineering discipline (*biomimetics*) that looks at the biological world and tries to use the lessons of nature to solve engineering problems. A common analogue is that between the hypodermic syringe and a mosquito (Figure 6.4).

Figure 6.3
Analogy of "walking" to generate ideas.

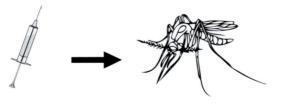

Figure 6.4
Analogue between a hypodermic syringe and a mosquito.

6.4.4 *Brainstorming*

Nowadays "brainstorming" is not a politically correct name for a really useful exercise, but I have yet to find an adequate alternative. Some of my clinical colleagues have a coarser name; they call it *brain dumping*. In effect it is simply a collection of people in an applicable space just spouting ideas off the top of their heads. As with all group activities no one's idea is wrong and there must be someone jotting down the ideas as they develop. Sometimes this type of meeting can be totally anarchic and difficult to manage (herding cats is a good analogy); sometimes the ideas flow like water and it is hard to keep up. To get the latter pick a good venue, have good icebreaking activities, and make sure that all participants feel equal and included.

Some basic rules to brainstorming are the following:

- Quantity is good: don't worry about quality at this stage, capture as many ideas as you can.
- Criticism is bad: do not let members laugh at, criticize, or ridicule another member's ideas.
- Creative thought is good: let members go off at tangents but do not let them get distracted.
- Combinations are good: if two ideas merge into one then so be it.

One last trick, make sure you look at *one* thing at a time; it is very easy, while brainstorming, to get distracted.

6.4.5 *Discretizing*

To help with any of the activities it is often beneficial to cut the overall idea into smaller parts. "Discretizing" comes from finite element analysis and literally means "cutting into discrete parts." For example, if we were getting ideas for testing blood samples we may discretize the whole system into four parts (Table 6.2).

Table 6.2: Example of Discretizing

Main Function	Discretized Functions
Testing a blood sample	i) collecting the blood sample; ii) transferring the blood sample; iii) analyzing the blood sample; iv) disposing of the blood sample.

6.4.6 Morphological Analysis

This in essence is an extension of discretizing but instead displays the specification in tabular form (as illustrated in Table 6.3).

The left-hand column is taken from the specification and/or the discretizarion described previously. The top row simply allows you to predetermine a maximum number of permissible solutions. The subsequent spaces are to be completed, but in each box a potential solution for that feature is inserted. In this way an NxN table is completed that hopefully provides every solution to every problem. The completion of this table is best illustrated by example. Using the blood sample analysis system described earlier we can build a picture (Table 6.4).

The basis of morphological analysis is to look for paths: How do the individual solutions "morph" into the bigger single solution? As we can see from Table 6.5, depending on which solution is picked for feature 1 a solution to feature 2 follows; it also enables us to see which solutions are "more versatile." An overall path, or a number of overall paths, should lead from the top to the bottom. If a path is not possible, or is incomplete, then another solution needs to be thought of and this can often provide focus. Note that this table also makes sure your *outputs* are compared with *inputs*!

Table 6.3: Example Morphological Chart

Specification feature	Solution 1	Solution 2	Solution 3	...	Solution N
1 .... N					

Table 6.4: Example Morphological Table for Blood Sample Analysis

Specification feature	Solution 1	Solution 2	Solution 3	Solution 4	Solution N
1. Collect sample	Syringe	Pin prick	Incision		
2. Transport sample	Sample bottle	Absorbent sheet	Slide	On instrument in 1	Direct to device (3)
3. Analyze sample	...				
4. Dispose of sample	...				

Table 6.5: Example of Morphological Analysis

Specification feature	Solution 1	Solution 2	Solution 3	Solution 4	Solution N
1. Collect sample	Syringe	Pin prick	Incision		
2. Transport sample	Sample Bottle	Absorbent sheet	Slide	On instrument in 1	Direct to device (3)
3. Analyze sample	.....				
4. Dispose of sample	.....				

6.4.7 Research

Although this has been left to last, it should not be your last point of call. Remember the Newton quote "on the shoulders of giants"? Why reinvent the wheel if you do not have to. Quite often the solution is, quite literally, staring you in the face. But where do you start?

The first place is lapsed patents. If a patent has lapsed then it is free to be used. If you have the money, then you are also able to buy or license active patents. Every government has a patent office, and every national patent office has a patent search engine, hence all you need is access to the web to perform this simple search.

The second place to look is a company catalog (and trade literature). Often there is something sitting on a shelf that you can buy that is totally fit for your purpose. Because you're buying it off the shelf you have no development costs, and it will probably be cheaper in the long run. Do not be afraid of buying technology; however make sure that what you have found is "*medical use compliant*"!

The third place to look is scientific literature. Quite often a research project in a university will present something in a paper that solves your problem. They may or may not have it protected by a patent. If they haven't then it is public domain and you can use it; if it is protected you will need to contact the University IP office to discuss licensing. However be warned…University IP offices are not fast movers so do not expect a quick answer.

6.4.8 We Have Ideas!

The whole reason for Section 6.4 was to enable you to generate ideas, and lots of them. Generating ideas, if done sensibly, is relatively easy. Generating lots of ideas can be quite hard, and every now and then generating just one idea becomes tediously slow. However, using one of the tricks listed above should "oil the wheels of creativity." The hard part is yet to come, which is how do we pick the best idea? That is the topic for the next section.

6.5 Selecting Concepts and Ideas

While the generation of ideas is, in the terms of cricket, a *free hit*[1] , the selection of the most appropriate idea is more formal and often more difficult. To this end I will present a few tools that have stood the test of time.

6.5.1 Morphological Analysis

Table 6.4 illustrated the morphological table including all paths. However, it should be apparent that at least one path would "fit the bill." Table 6.6 illustrates this for a device to monitor the temperature of a patient, constantly.

The black arrows show a number of paths that could be taken. However, the optimal path is highlighted in gray. In order to fulfill the audit trail the reasons for selecting this gray path should be given. They may be quite subjective, as design choices often are, but they must be written down. So, for example, the data transmission may have been either Bluetooth or ZigBee, but ZigBee was chosen because more than one temperature probe can be monitored. Or it could have been ZigBee because it would be the only thermometer on the market using this technology. The best way to log these decisions is to annotate the diagram (if possible), or to number the arrows and have a separate sheet with the arrow number and reasons for the choice given (as illustrated in Table 6.7).

As you can see this table could be very verbose and, hence, hard to complete for very large projects. The next option is a numeric-based system.

6.5.2 Criteria Assessment

You will have spent a lot of time developing a specification hence it makes sense to use it. If you have discretized your project then you will have split and reformed the PDS too. This

Table 6.6: Example of Morphological Analysis to Select an Ideal Solution

Specification Feature	Solution 1	Solution 2	Solution 3	Solution 4	Solution 5
1. Measure temperature	Mercury thermometer	Infrared camera	Thermocouple	Temperature sensitive gels	Integrated circuit chip
2. Transmit data	Wires	Bluetooth	ZigBee	Radio	Infrared
3. Collect data	PC	Smartphone	Pad	Chart plotter	
4. Plot data	Printer	Plotter	Screen	Chart plot	

---

[1] A free hit occurs in some games of cricket to stop a bowler from constantly bowling illegal balls. If an illegal ball is given then the next ball is a "free hit" and the batsman cannot be given out. Hence it is a relatively easy task!

Table 6.7: Example of Morphological Analysis to Select an Ideal Solution with Reasons

Specification Feature	Solution 1	Solution 2	Solution 3	Solution 4	Solution 5
1. Measure temperature	Mercury thermometer	Infrared camera	Thermocouple	Temperature sensitive gels	Integrated circuit chip
2. Transmit data	Wires	Bluetooth	ZigBee	Radio	Infrared
3. Collect data	PC	Smartphone	Pad	Chart plotter	
4. Plot data	Printer	Plotter	Screen	Chart plot	

Arrow	Reason
1	Integrated chip selected as it measures T using low voltage technology and Bluetooth or ZigBee transmission is integrated. Mercury therm rejected for obvious reasons! Infrared camera is possible but too expensive. Thermocouples are a possibility but require specialist systems to connect to. Temperature-sensitive inks would need a camera to capture changes. ZigBee transmission selected as this allows for more than one T probe to be monitored at any time without the need for switching. Price is within budget.
2	Data collection is to be a PC as all wards have one at main reception desk. It cannot be assumed that they will have tablet/pad technology.
3	Integration of ZigBee receiver into a PC is commonplace and can be bought off the shelf.

method of assessment simply grades the potential solutions versus the specification itself. Hence the table becomes something like Table 6.8.

The aim here is to assess how well a solution meets the requirements laid down by the specification. There is much debate over the grading system. Some use 0–3, with 0 being not at all, 1 not very good, 3 very good, and 2 average. Some use 0–10, some use 0–5. In the end it doesn't matter so long as you are consistent. Personally I prefer 0–10 but going up in twos; 0 – not at all; 2 – a little; 4 – below average; 6 – above average; 8 – very good; 10 – perfectly. This then leaves the door open to finer grading, for example if you have three solutions with one on 6 and one on 8 but a third in between the two.

As an example let us examine the method of temperature measurement in more detail. The PDS has six main requirements; these are in the first column. The solutions are now inserted in the top row. The empty spaces are now ready for grading.

Table 6.8: Criteria Assessment Table

Specification item	Solution 1	Solution 2	.....	Solution N
PDS item #1	Numeric grade	Numeric grade		Numeric grade
PDS item #.	Numeric grade		Numeric grade	Numeric grade
...	...	...	...	...
PDS item #N	...	...	...	...
Total score	Σ↓	Σ↓	Σ↓	Σ↓

Table 6.9: Criteria Assessment for Measurement of Temperature

Specification Item	Mercury Thermometer	Infrared Camera	Thermocouple	Temperature Sensitive Gels	Integrated Circuit Chip
1. Measure temperature from 0–40°C					
2. Not toxic					
3. Reusable after suitable sterilization					
4. Low power consumption					
5. Low purchase cost					
6. Ability to transmit data					
Total Score					

Clearly the grading is subjective. However, one way of removing subjectivity is to pass the table to others in your design team and let them do the grading. You can then accumulate the data into one table (by adding the scores together), producing a more coherent result. After some analysis your table may look like Table 6.10. Note the last thing to do is sum the columns, with the highest score giving you your best solution.

The table illustrates two front-runners, but the totals are so close there is little to choose between them. There is one column with zeroes; the rule is any zeroes usually mean automatic elimination from the process. Normally this happens by default but you need to keep your eyes open.

6.5.3 Weighted Criteria Assessment

If, as in Table 6.10, the result is close or there is no clear winner, no champion, weighting the criteria can be very useful. Equally you can simply use it all of the time. Here

Table 6.10: Example of a completed criteria Assessment for Measurement of Temperature

Specification Item	Mercury Thermometer	Infrared Camera	Thermocouple	Temperature Sensitive Gels	Integrated Circuit Chip
1. Measure temperature from 0–40°C	10	10	10	10	10
2. Not toxic	0	10	8	7	9
3. Reusable after suitable sterilization	10	10	9	1	2
4. Low power consumption	10	1	4	10	8
5. Low purchase cost	0	1	5	8	7
6. Ability to transmit data	4	10	8	6	9
Total Score	34	42	44	42	45

the individual rows are weighted relative to their importance. For example there may be a specification requirement for the device to be blue, but compared with another requirement stating it should be nontoxic this is relatively minor. Hence to give the rows the importance they deserve it is common to give them a weighting in either decimal or percentage terms. The easiest way to do this is to rank them with the most important being ranked first, and the least important being last. Once again, the best way to obtain this ranking is to pass the list to a group and let them, individually, do the ranking. It is then relatively easy to assimilate the marks to obtain a syndicated, average ranking table. This is also a very good way to include customer/end-user input. To calculate the rank we use the formula

$$W1 = 1 - \frac{i - 0.5}{n} \tag{6.1}$$

Alternatively you can just attribute the weighting as

$$W2 = 100 \frac{n - i}{n - 1}\% \tag{6.2}$$

But this means the last one always has a zero ranking. Just a simple reversal of the rank (see W3) often works.

So for example the weighting of 6.9 may be as in Table 6.11.

Note that I have made items 4 and 6 equal at 4.5, which is midway between 4 and 5 (i.e., they share an equal rank). If there were three sharing it would be 4.33; four sharing 4.25, and so on. The next rank number is then back to sequence, i.e., 6 and not restarting at 5. Hence for

Table 6.11: Weighted Criteria Assessment for Measurement of Temperature

Specification Item (n = 6)	Rank (i)	W1	W2	W3
1. Measure temperature from 0–40°C	1	92	100	6
2. Not toxic	2	75	80	5
3. Reusable after suitable sterilization	6	8	0	1
4. Low power consumption	4.5	33	30	1.5
5. Low purchase cost	3	58	40	3
6. Ability to transmit data	4.5	33	30	1.5
Total Score				

Table 6.12: Weighted Assessment[*]

Specification Item	W1	Mercury Thermometer	Infrared Camera	Thermocouple	Temperature Sensitive Gels	Integrated circuit chip
1. Measure temperature from 0–40°C	92	9.2 (10)	9.2 (10)	9.2 (10)	9.2 (10)	9.2 (10)
2. Not toxic	75	0	7.5(10)	6(8)	5.25(7)	6.75(9)
3. Reusable after suitable sterilization	8	0.8(10)	0.8(10)	0.72(9)	0.08(1)	0.16(2)
4. Low power consumption	33	3.3(10)	0.33(1)	1.32(4)	3.3(10)	2.64(8)
5. Low purchase cost	58	0	0.58(1)	2.9(5)	4.64(8)	4.06(7)
6. Ability to transmit data	33	1.32(4)	3.3(10)	2.64(8)	1.98(6)	2.97(9)
Total Score		14.62(34)	21.71(42)	22.78(44)	24.45(42)	25.78(45)

*Numbers in brackets are original scores before weighting.

three sharing the sequence would be 4.33, 4.33, 4.33, 7 (instead of 4, 5, 6, 7). Supposed we used W1 as our criteria, how does that affect our choice of champion?

Once again the mercury thermometer falls out because of the zero, but its score is so low it has to fail. Also, the front-runner is now clear (Table 6.12).

6.6 Summary

This chapter has introduced you to the concepts of ideas generation and methods of selection. We have seen how important the working environment is, and how important those you invite to help you become. Several tools have been illustrated to help you generate ideas, concepts, and solutions. To become proficient you must practice. Do not expect to be able to walk into your first brainstorming session and for it to work smoothly; it takes many tries before you hit on your formula.

Once we have our ideas we saw three methods of selection. By far the most powerful is to use weighted criteria. We should recognize that this single table ensures that your FDA/MDD auditors see that you are comparing inputs with outputs!

References

Dym, C. L., & Little, P. (2000). *Engineering design – a project based introduction*. Chichester: J Wiley and Sons Ltd.

Grossman, S. R. (1988). *Innovation, Inc.: Unlocking creativity in the workplace*. Wordware Publishing.

Hurst, K. (1999). *Engineering design principles*. London: Arnold Publishers.

Lloyd, P. (2011). *Creative space*, <http://www.gocreate.com/articles/creative-space.htm> cited 12.09.11.

Michalko, M. (2001). *Cracking creativity: The secrets of creative genius*. Ten Speed Press.

Ulrich, K. T., & Eppinger, S. D. (2003). *Product design and development*. McGraw Hill.

## CHAPTER 7

Quality in Design

7.1 Introduction

You are probably bored of me quoting ISO 13485, MDD, and the FDA. Unfortunately they are very important to us. One of the main reasons for having our own ISO for medical devices companies was that the totality of the ISO 9000 family did not sit comfortably with our discipline. However it is, effectively, a sibling of the ISO 9000 family and as such is concerned with ensuring quality. While we have formulated procedures to meet the ISO requirements, they do not themselves ensure that a *quality item* has been designed. We can produce a brilliant paper trail showing that we have met ISO 13485 – however it is the detail in the paper trail that actually determines the quality of the device. In this chapter we will look at design tools specifically developed to makes sure your design is an *optimum design*.

I am sorry but this chapter has some mathematical analyses. It was inevitable that you had to get your calculator out at some point. However most of the tools I describe can be undertaken in a spreadsheet, so maybe it's time to invest in a personal laptop and a copy of a spreadsheet program. It may even be time to seek out the dedicated software that is available or to team up with someone (such as a university) who has a copy. In this chapter we shall be examining specific tools for design activities that promote quality. Specifically we shall examine optimization, design of experiments (2^k factorial), House of Quality, FMEA, D4X, and 6σ.

7.2 Optimization

One of the main uses for optimization is to minimize mass. If you consider the aircraft industry then it is pointless having a cargo plane that can only carry its own body weight. To be optimal the body weight must be a minimum so that its payload is a maximum. This is the point of optimization – you need to have an "objective" to achieve; this could be maximum power, it could be minimum weight, it could be maximum volume. You then have to manipulate your design to see if you can achieve the optimum. This often involves the manipulation of numerous mathematical models – they need not be complex but there will always be more than one.

Optimization is not new – it is the basis of evolution. Our human bodies have the most wonderful optimization system built-in; it is called *bone*. Your bones remodel all of the time based on the load they carry; in fact your skeleton is renewed every two years.

Medical Device Design.
DOI: http://dx.doi.org/10.1016/B978-0-12-391942-7.00007-6

If your bone is heavily loaded, say due to sport, your bones will gain mass. If your bone is lightly loaded, say due to being in space and weightless, your bones will lose mass. They are constantly changing and trying to achieve an optimal solution. In bone this is called *remodeling*.

There is a term for the ultimate objective you are trying to achieve; this is called the *objective function*. There may be more than one. It is a mathematical expression used to model your design. It is usually written in the form of an expression

$$f_o(A,B,C) = f(A,B,C)$$

(where f_o is the objective function and f is any mathematical function of the variables A, B, and C). Or the objective function of parameters A, B, and C is defined by the equation on the right-hand side.

The simplest form of optimization is *linear*. Often in mathematics you will see the term *linear programming*. The best way to visualize this is as a graph of two straight lines. Suppose we have a system where the objective function is

$$f_o = 3x + 4y$$
$$\int where$$
$$0 < x \le 5$$
$$and$$
$$y \ge 4$$

and we need to minimize f_o. Figure 7.1 illustrates the objective function (values of 16, 24, and 32) and the constraints. The constraints mean that the solution can only lie in the shaded region. By inspection the minimum lies in the bottom left-hand corner when $x = 0$ and $y = 4$, giving $f_o = 16$. All other values of x and y either lie outside of the constraints (outside the design space) or result in values greater than the minimum (16).

Consider a cylinder of diameter D and length L made from steel plate of thickness $t = 5\,mm$ then we have two possible objective functions: mass of the cylinder itself and the volume it contains (Figure 7.2).

The mass of the cylinder is given by

$$f_o(D,L)_1 = \left[\pi DLt + 2\left(\frac{\pi D^2}{4} \right) t \right] \rho$$

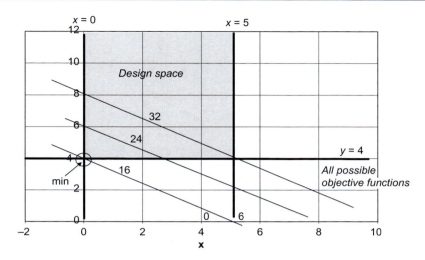

Figure 7.1
Graph of objective function and its solution.

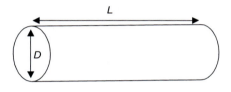

Figure 7.2
Cylinder model.

The volume it contains is given by

$$f_o(D,L)_2 = \frac{\pi D^2}{4} L$$

The next thing is to determine the *constraints*. What are the minimum and maximum values of the parameters? Which, if any, of the parameters are fixed? If we go back to our design concepts these limits give us the boundaries of the design space. We will then be looking for an objective: Are we trying to minimize mass for a fixed volume? Are we trying to maximize volume for a fixed mass? We need to know what we are looking for, but we also need to know a tolerance. We will never find an exact solution but we may find one if we state we are looking to find a solution where the mass is minimized but the volume should be 0.995–1.005 liter.

In graphical terms this is like plotting the design space as a surface and using the objective function to determine a solution, as illustrated in Figure 7.3.

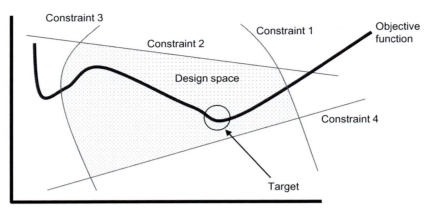

Figure 7.3
Illustration of optimization.

CASE STUDY 7.1

The diameter of a hollow cylinder with closed ends may not exceed 100 mm and its length may not exceed 200 mm but must not be shorter than 50 mm. The thickness of the material is fixed at 5 mm. The target volume is 0.995–1.005 liter and the mass of the cylinder is to be minimized. Determine the values of cylinder diameter and length that satisfy these criteria. Our optimization problem is written as follows:

Objective function:

cylinder mass

$$f_o(D,L)_1 = \left[\pi DLt + 2\left(\frac{\pi D^2}{4} \right)t \right]\rho$$

(where r = density of steel at 7850 kg/m^3)

Constraints:

cylinder volume

$$f_o(D,L)_2 = \frac{\pi D^2}{4}L$$

$$t = 0.005\,\text{m}$$

$$0 < D < 0.1\,\text{m}$$

$$0.05 < L < 0.2\,\text{m}$$

Target:
$$0.995 < V < 1.005 \text{ liter}$$

M minimized

If we create a table of possible combinations then we can start to build a picture. If we start with four values of diameter we will have corresponding lengths (from the volume criteria/objective function), and as a consequence corresponding cylinder masses (cylinder mass objective function).

Table 7.1 shows that as the diameter of the cylinder increases the length required decreases (as expected). A quick scan of the table illustrates that the optimum solution is to have a diameter of 50 mm and a length of 50.7–51.2 mm, which yields the correct volume and the minimum mass of about 0.47 kg.

Table 7.1: Optimization Table

Diameter D	Volume V	Length L	Cylinder Mass M
(mm)	(liter)	(mm)	(kg)
25	0.995	202.8	0.66
25	1.005	204.8	0.67
50	0.995	50.7	0.47
50	1.005	51.2	0.47
75	0.995	22.5	0.55
75	1.005	22.8	0.56
100	0.995	12.7	0.77
100	1.005	12.8	0.77

Clearly this was a very simple problem. But it illustrates the power of optimization. For more complex systems this method will not work and you will need to use one of the many techniques available (such as linear programming, Routh–Hurwitz, Monte Carlo method, etc.). Most modern computer-based mathematics programs contain optimization routines, but you must define the parameters. There is no need to purchase a program – many open-source programs can be found on the web. If you have access to Microsoft Office® (Microsoft, 2011) then you have their optimization routine called *solver* (under Tools – Solver in Excel). Use of this routine reveals the result shown in Table 7.2.

Most computer-aided design (CAD) packages, such as Solidworks® and ProEngineer®, come with built-in analysis that enables you to perform design optimization from the very solid models you are drawing. There is really no excuse not to undertake some form of optimization.

Table 7.2: Optimal Result Obtained Using Microsoft Excel® Solver Routine

Diameter (mm)	Length (mm)	Mass (kg)
50.32	50.32	0.47

7.3 Design of Experiments/2^k Factorial Experiments

Sometimes it is difficult to ascertain which parameter in your design is the most important. Equally sometimes you need to ascertain which parameter determines the quality of your design. So, for example, if we were to examine the performance of a sphygmomanometer (used to measure blood pressure) we would have many variables to consider, but which are insignificant? Which have a detrimental effect on performance? Which have a beneficial effect on performance? When the Japanese were tussling with improving quality an engineer named Taguchi realized that it was important to design out problems. Hence he needed a simple experiment to determine which parameter of a design has the greatest effect on quality, and if it is detrimental get rid of it. To do this he invented *factorial experiments*. There are whole textbooks on this subject so I can only give an introduction. However, the tool I am about to share can be used for most design problems. It should be noted that you do not need to have complex mathematical models for this to work; testing real things is possible too.

Consider the injection molding of a syringe body. The variables we have on the injection molding machine are T = temperature of the injected plastic, P = the pressure of the injected plastic, and Tm = the mould temperature. Now this is where Taguchi was clever; instead of looking at a whole range of values he proposed we should only look at maximum and minimum values of all parameters. So if we could set the mould temperature to be anywhere between −5 °C and 20 °C these would be the extremes. As there are three variables (T, P, and Tm) that can only be set at two values, the total number of experiments required is 8 (2^3). Hence for any system with k variables the total number of experiments is 2^k. To design the experiment we use −1 to signify a minimum, and +1 to signify the maximum. The design of the experiment is simple (see Table 7.3).

It is pretty obvious that beyond four variables 2^k experiment design is time-consuming and, if you are doing real experiments, costly. If, however, you are doing numerical-based models then the only cost is time; and this can be reduced by using a computer-based model. I have further highlighted the subsets for three variables and two variables for your information. For systems with more than variables it is common to use $^{k-n}$ experiment design, but that is beyond the scope of this text.

Now let us return to our original example with three variables; how does this affect Table 7.3?

Table 7.3: Experiment Design for a Four-Variable System

Experiment	Variable 1 (X1)	Variable 2 (X2)	Variable 3 (X3)	Variable 4 (X4)
1	+1	+1	+1	+1
2	+1	+1	+1	−1
3	+1	+1	−1	+1
4	+1	+1	−1	−1
5	+1	−1	+1	+1
6	+1	−1	+1	−1
7	+1	−1	−1	+1
8	+1	−1	−1	−1
9	−1	+1	+1	+1
10	−1	+1	+1	−1
11	−1	+1	−1	+1
12	−1	+1	−1	−1
13	−1	−1	+1	+1
14	−1	−1	+1	−1
15	−1	−1	−1	+1
16	−1	−1	−1	−1

Table 7.4: Experiment Design with Settings

Experiment	Variable 1 (X1)	Variable 2 (X2)	Variable 3 (X3)	T	P	Tm
1	+1	+1	+1	Max	Max	Max
2	+1	+1	−1	Max	Max	Min
3	+1	−1	+1	Max	Min	Max
4	+1	−1	−1	Max	Min	Min
5	−1	+1	+1	Min	Max	Max
6	−1	+1	−1	Min	Max	Min
7	−1	−1	+1	Min	Min	Max
8	−1	−1	−1	Min	Min	Min

Now we come to the interesting bit, we do the experiments. Using the table we add the settings values; they could be just Max and Min, or they could be actual numbers, it matters not (Table 7.4). We have eight experiments to conduct, but we must measure something. We could measure quantitative numbers such as weight, surface finish or number of surface defects. Equally we could be quite subjective and grade how it feels or looks (from 1–10 say). Again it matters not so long as we have a measure of quality described by a number. Before we do the experiment we mess up the data using a random selection for the experiment order; this eliminates any order effects. The final column is the results from the experiment (in Table 7.5 perceived quality is rated from 0–5 where 5 is excellent and 0 is awful). It is a really good idea to get a third party to do the experiments. If you do this you will have just made the test more statistically relevant by removing your influence from the results.

Table 7.5: Experiment Randomized and Completed

Random Run Number	Experiment	Variable 1 (X1)	Variable 2 (X2)	Variable 3 (X3)	T	P	Tm	Result (Q)
7	1	+1	+1	+1	Max	Max	Max	4
6	2	+1	+1	−1	Max	Max	Min	1
4	3	+1	−1	+1	Max	Min	Max	2
5	4	+1	−1	−1	Max	Min	Min	4
3	5	−1	+1	+1	Min	Max	Max	2
8	6	−1	+1	−1	Min	Max	Min	3
2	7	−1	−1	+1	Min	Min	Max	4
1	8	−1	−1	−1	Min	Min	Min	4

We analyze the results taking each +1 and −1 of each variable in turn in relation to the result Q. So for variable X1 the +1 results were experiments 1–4. For variable X2 the +1 results were experiment 1, 2, 5, and 6. We analyze them in this fashion:

Average +Q = Sum (+1 values of Q)/Number of results
Average −Q = Sum (−1 values of Q)/Number of results
Variance of Q = (Average +Q) − (Average −Q)

Or, more elegantly

$$E_1 = \frac{\sum Q(+1)}{2^{k-1}}$$

$$E_2 = \frac{\sum Q(-1)}{2^{k-1}} \qquad (7.1)$$

$$E = E_1 - E_2$$

For variable X1 this would be as shown in Equation 7.2 and Table 7.7.

Maximum   *Minimum*

$$E_1 = (1.37 + 1.42 + 1.11 + 1.4)/4 = 1.32 \quad E_2 = (0.92 + 1.84 + 1 + 1.37)/4 = 1.28 \qquad (7.2)$$

Overall

$$E = 1.32 - 1.28 = 0.04$$

Similarly we can determine the values for the system for variables X2 and X3 (Table 7.7). Or we can plot these on an effect graph (Figure 7.4).

Both Table 7.7 and Figure 7.4 illustrate that the variable X3 has the highest effect – its slope is negative which means the effect is decreasing. X1 and X2 seem to have a similar effect.

Table 7.6: Subset of Table 7.5 to Highlight Results for Variable X1

Random Run Number	Experiment	Variable 1 (X1)	Result (Q)
7	1	+1	1.37
6	2	+1	1.42
4	3	+1	1.11
5	4	+1	1.40
3	5	−1	0.92
8	6	−1	1.84
2	7	−1	1.00
1	8	−1	1.37

Table 7.7: Effects Analysis for Variables X1, X2, and X3

	Min	Max	
X1	1.28	1.32	0.04
X2	1.22	1.28	0.06
X3	1.51	1.10	−0.41

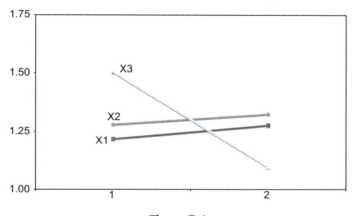

Figure 7.4
Effect diagram for parameters X1, X2, and X3.

However, we now need to examine how X1 interacts with X2, etc. This requires *no more experiments*! We simply analyze the data using a table generated from the previous results. So the value for interaction between X1 and X2 is the value in the X1 column multiplied by that in the X2 column (not the results column – the +1, −1 columns). As we only have +1 and −1 to worry about the answers can only be +1 and −1. Since we have three variables and they all have to interact with one another, we have to determine the number of new columns:

$$\text{No. of interactions} = N! - (N - 1)! \qquad (7.3)$$

which for three variables becomes

$$3! - 2! = 4$$

These interactions are X1.X2, X1.X3, X2.X3, and X1.X2.X3.

Hence our table is modified as shown in Table 7.8.

The analysis of the results uses Equations (7.2) again – but we need no more experiments! Our new analysis table is shown in Table 7.9.

From Table 7.9 we can see that the dominant effect is created by all three parameters in combination. However, we do not know if this is a statistically significant variation, or if it is just due to simple random variance. More often than not the variations found are simply due to random variations (and these in turn are due to tolerances). There is little we can do with these, but the information does help us to decide where to tighten tolerances and where it is possible to slacken them.

Table 7.8: Inclusion of Interactions into Analysis Table

Random Run Number	Experiment	X1	X2	X3	T	P	Tm	X1.X2	X1.X3	X2.X3	X1.X2.X3	Result (Q)
7	1	+1	+1	+1	Max	Max	Max	+1	+1	+1	+1	1.37
6	2	+1	+1	−1	Max	Max	Min	+1	−1	+1	−1	1.42
4	3	+1	−1	+1	Max	Min	Max	−1	+1	−1	−1	1.11
5	4	+1	−1	−1	Max	Min	Min	−1	−1	+1	+1	1.40
3	5	−1	+1	+1	Min	Max	Max	−1	−1	+1	−1	0.92
8	6	−1	+1	−1	Min	Max	Min	−1	+1	−1	+1	1.84
2	7	−1	−1	+1	Min	Min	Max	+1	−1	−1	+1	1.00
1	8	−1	−1	−1	Min	Min	Min	+1	+1	+1	−1	1.37

Table 7.9: Effects of Parameters (Including Interactions) Ranked in Order of E Ascending

	Max	Min	E
X3	1.0979	1.5063	−0.4084
X1X2X3	1.2037	1.4004	−0.1967
X1X3	1.3139	1.4196	−0.1057
X1X2	1.3139	1.2903	0.0236
X1	1.3236	1.2806	0.0431
X2	1.2764	1.2178	0.0586
X2X3	1.3416	1.2625	0.0791

To determine their respective significance we undertake a simple statistical analysis. First we rank the parameters in order of variation, starting with the most negative and ending with the most positive. Once ranked we determine the probability of that effect being in that position using

$$P_i = \frac{i - 0.5}{2^k - 1} \tag{7.4}$$

The value Z is determined from standard tables (see Appendix B, Table B.1 – you should try to replicate the Z column in Table 7.10 for yourself).

If, in your table, you have items with the same value of E then you must follow the division rule as seen earlier. So, for example, if X1X2 and X1 had identical values of E, the rank list would be 1, 2, 3, 4.5, 4.5, 6, 7.

The next part of the analysis is to plot a graph of effect E versus Z, as in Figure 7.5. Most of the points lie on the normal straight line (this will be discussed more later). However X3 stands out like a sore thumb – as does X2X3. They are clearly different from the rest.

Note that the normal line goes through zero; this is obtained using a best-fit line and forcing 0-0. The graph indicates a "wrong" best fit which goes through the points, but not through 0-0; a common error made by those doing their first 2^k experiments. The best way of achieving this is to plot the graph (as an x-y scatter graph) in a spreadsheet program, then add a linear trend line with the "crosses at zero" option flagged. But be careful, as those with severe outliers will distort your best-fit line! Figure 7.6 demonstrates this where there are obvious outliers from the straight line. These are the significant effects; all others are random aberrations over which you have little control, apart from tightening tolerances.

These two points are outliers whose effects are statistically significant, or whose effects are due to the change in parameter and not simply due to random variation. Thus the same argument applies to X3 and X2X3 in Figure 7.5.

Table 7.10: Normalized Scores for the Parameters' Effects

	E	Rank	P	Z
X3	−0.4084	1	0.07	−1.465
X1X2X3	−0.1967	2	0.21	−0.792
X1X3	−0.1057	3	0.36	−0.565
X1X2	0.0236	4	0.50	0
X1	0.0431	5	0.64	0.565
X2	0.0586	6	0.79	0.792
X2X3	0.0791	7	0.93	1.465

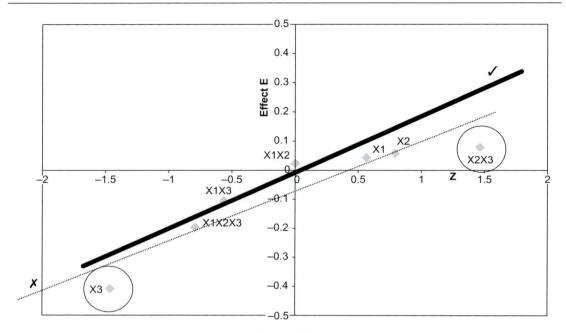

Figure 7.5
Normalized plot of all effects.

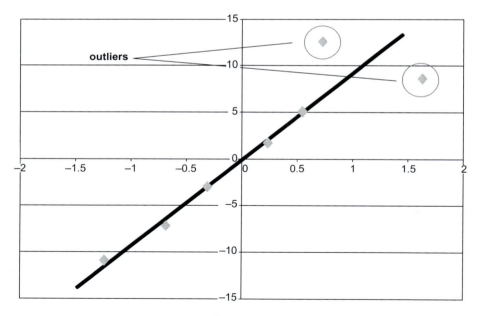

Figure 7.6
More common form of normalized plot.

The next question is: What do we do with this data? There are two interpretations. The first concerns minimizing manufacturing errors. If you want to make your quality control more robust then you must examine the outliers and see how to best control them or remove them from the design. Clearly, if you are able, then these should be tightly toleranced. All others have a small effect but your level of control need not be so high, if at all.

The second way of looking at the data is if you are trying to maximize the effect; now the data tells you which parameters are the most important and hence which you should vary to improve your outcome.

As you can see, 2^k experimentation (or "design of experiments") is an extremely powerful tool. I have only presented the most basic of introductions; if you wish to understand more then a good text is Montgomery (2001). If you would like to play with this method conduct a web search for "Taguchi Paper Helicopter" and you will find links to a great example of "design of experiments."

7.4 House of Quality

We have appreciated the importance of customer/end-user input into design. But how do we know that we have really taken them into account, and how can we use this knowledge to differentiate ourselves in the marketplace? One of the most valuable tools is the House of Quality (or HoQ as I shall now term it). Figure 7.7 is a schematic of an HoQ structure.

The HoQ is split into zones. The first "room" is customer requirements; in this room the individual requirements, as detailed by your customers and end-users are tabulated as

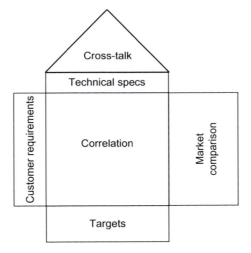

Figure 7.7
Typical House of Quality structure.

individual rows. The second room – "technical specifications" – is *your* items from the PDS that you use to define your device. These are tabulated as individual columns. This table creates a square zone where the customer requirements and technical specifications are cross-correlated. The intention of this central room is to ensure that you have at least one technical specification for every customer need. So, for example, the customer may have asked for the device to be blue, hence you should have a technical specification that defines color in some way (say, a coating). The foundation of the house is the targets for your technical specifications. So if one of your items was to define the power utilization of your device you may have a target to not exceed 4.5 kW. The lean-to, to the right of the house, is used to look at how well your competitors meet the customer requirements. Both the foundation and the lean-to enable you to define benchmarks for your device that you can use in marketing and to influence development. The last part is the roof; this is where you examine the interaction between your technical specifications. For example, you may have a requirement to maximize component strength and another to minimize component weight. Clearly both interact with one another; the roof helps you to decide how they interact and in which direction you want them to go.

The best way to see how the HoQ works is to examine one. Figure 7.8 illustrates a completed HoQ for a piece of clinical software to help examine x-rays on a mobile smartphone.

Clearly you should appreciate that Figure 7.8 is a foreshortened example, however it demonstrates the principle. Note that the items in the technical requirements are very much related to how to do it, how much is needed, etc. The customer requirements tend to be more wooly and less specific, but not necessarily so. The central room of the house allows you to correlate customer requirements against your technical specification. You should examine each box in turn and insert a bold cross if it is strongly related and a normal cross if there is a weak correlation. If there is not a correlation add nothing. The important note here is no horizontal row should be empty. If it is empty then you have a technical specification missing! Once you are happy with the central room you can go to any section of the house, in any order.

So if we continue along the rows we enter the lean-to. In here we enter real numbers. If there are competitors we create columns for each competitor and one for our new device. In here we grade how well the competitors and our new device fit the customer requirements: 0 for not at all and 100% for fully meeting the requirements. This area enables us to examine where we can make improvements to create market differentiation. It can also help us to decide what number should go into the basement. So, for example, if our competitors can only flip a photo about a vertical axis we can make an improvement by flipping about a horizontal axis too.

Let us now go into the basement. In here we enter targets. Hence, although in the specification it says all platforms, in here we specify which platforms. In the technical requirements it says

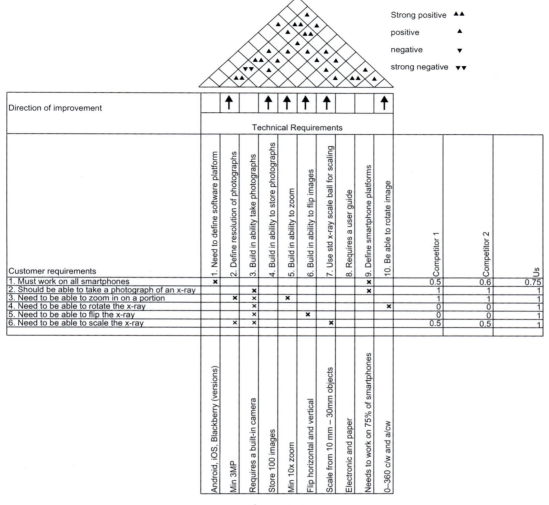

Figure 7.8
Example HoQ for smartphone application.

we should store photographs, in here we specify how many (or the minimum). Once again we can use these columns to compare against our competitors. We can also add a row which demonstrates the degree of difficulty associated with the target (again 0 being no issue at all and 100% being very, very, very hard).

Now we move into the roof or, if you wish, the attic. We pass through the first floor landing, direction of improvement. Here we look at the technical requirements and decide which way they need to go. For example we have an arrow pointing up for the storage of photographs as we would like to save as many as possible. If we needed to reduce something, say power usage, the arrow would point down. This informs you of how the targets are to be met.

At last we come to the attic. This layout enables you to compare each technical specification item against another. It works in the same way as a football chart showing who is playing who over a season. In these boxes we add one of four symbols. Double arrows up mean that the two are correlated and the change is strong and improving. A single arrow up means the change is weak. No symbol means there is no link. A single downward arrow means the change is not improving but is detrimental; a double downward arrow is strongly detrimental. The roof is important as it is the first time you look at the interaction of your decisions. The best way to exemplify this is with an automobile engine. If we have a target to have low consumption (i.e., more miles per gallon), but also a target to have an engine with more power output, you can see the two are correlated but increasing the power output will obviously decrease the miles per gallon. Hence it forces you to look at ways of turning the arrows upward or at least making them blank.

As with previous sections this is only a taste of HoQ. If you want to find out more about this subject area then you can find a liberal amount of books on the subject under the title QFD (Quality Function Deployment); there is also a liberal amount of good information on the web.

7.5 Failure Mode and Effect Analysis (FMEA)

This is one of the most valuable tools a designer can have in their toolbox. It is, literally, the place to imagine all the nasty things that can go wrong with your design and then to make them unlikely. Some people, incorrectly, think of it as risk analysis…we are not assessing risk (that comes later in this book). We are assessing how the device will fail, what will cause it, what the effect will be, and whether the effect is detrimental or not; and then using this analysis to improve our design (at the design stage!). It should not be underestimated that we are talking about foresight: a poor design that leads to a product recall is all about hindsight…a good designer never relies on hindsight! From now on do not let me hear you utter the phrase "*Well in hindsight we….*"; a good FMEA will remove this phrase from your phrasebook.

As an example let us consider the humble wheelchair. We may imagine many failure modes but let us imagine a person sitting in the chair, the fabric base ripping open and the person falling to the floor. The effects are obvious and traumatic; the failure is obvious – it ripped. However the causes are manyfold. One cause may be that the person was too heavy for the chair; another may be that the fabric was not strong enough to begin with; a third could be that the material was already ripped from bad treatment by the previous user. All three can exist but what do we do as a designer? We need to examine all the failures and try to design them out. If we are unable to design them out then we need to build limits into our device's usage.

FMEA was invented to accommodate this type of analysis. It is a simple table. It is not an assessment of risk, it is a tool to help you identify potential design faults and to help you concentrate on areas where your design could be deemed to be weak.

Table 7.11 illustrates an example of an FMEA chart. It has 11 columns, with the last four being repeats, but all will become clear.

Table 7.11 suggests a typical layout for an FMEA analysis. The first column is simply the number of the row (this helps later if you need to refer to it). The second column is the failure

Table 7.11: Example FMEA Template adapted from BS EN 60812 (BSI,2006)

Number	Failure	Effect	Cause	Rating				Remedial				
				Severity	Occurrence	Detection	RPN	Action	Severity	Occurrence	Detection	RPN
#	Text	Text	Text	S	O	D	RPN = S.O.D	Text	S*	O*	D*	RPN = S*.O*.D*

(b)		
S	O	D
1 = Inconvenience 2 = Temporary injury not requiring attention 3 = Injury requiring minor attention 4 = Permanent impairment or life-threatening injury 5 = Death	1 - 1/1000,000 2 - 1/100,000 3 - 1/10,000 4 - 1/1000 5 - 1/1000	1 = Detectable by anyone 3 = Detectable by skilled person 5 = Undetectable except by a rare highly skilled person (you can 1,2,3,4 5; these have been omitted for clarity of presentation)

(c)
S
1 = Inconvenience 2 = Temporary injury not requiring attention; can lead to a complaint 3 = Injury requiring minor attention; delay to procedure of < 30mins; could lead to a series of complaints 4 = Permanent impairment: life-threatening injury; potential for a product recall 5 = Death; cancellation of procedure; definite product recall

(d)
O
1 = One occurrence in 3 to 5 years or ½ in 1,000,000,000 uses 2 = One occurrence per year, or six in 100,000 uses 3 = One occurrence every 3 months or 5 in 1000 uses 4 = One occurrence per week or 5 or more in 100 uses 5 = More than one occurrence per day or more than 3 in 10 uses

mode – in this box you need to describe the imagined failure, in detail. The third column enables you to write down the effect of this failure mode. The fourth column enables you to determine the cause of the failure mode. As stated earlier there may be more than one cause for any failure mode; all have to be examined.

The next four columns enable you to assess the degree of failure. Some failures are inconsequential; others are really, really bad. Completing this table demonstrates that you have, at least, thought about them. The first of these ratings columns is "severity" (see Table 7.11(b)). An example rating is taken direct from the standard associated with risk analysis (ISO 14971:2009). The worst severity is death (given a 5); the least is a minor inconvenience (given a 1) – there should never be a zero. Note that the failure of your device may not have caused the death – your device may have failed and have a knock on effect to something it interacts with, i.e., a systemic failure, but you *must* consider this too. However Table 7.11(b) is only from an injury viewpoint; you can also damage your product's reputation. Consider the situation where a surgeon is in the OR using your device and it locks up so that it is unworkable and they have to cancel the operation. Would they ever use your equipment again? NO! To your company this failure is as bad as a death. As we are dealing with a medical device it makes sense to relate the effect to the effect on the patient, the person using it, etc. However we must not forget the effect on the company's "brand." Hence while *death* is clearly a bad outcome, for our company's brand a *product recall* is equally bad (I know this sounds crass, but it is a truism). Hence, unlike the later risk analysis, this FMEA covers any deleterious effect. Hence Table 7.11(c) incorporates severity that also affects your company's reputation in the marketplace so this table should be used…equally you can develop your own.

The next column in Table 7.11(b) is more problematic as it all depends on how many of your devices there are and how often they are used. So while I have suggested some guidelines there may be others; for example you could decide on the occurrence being based on the number of times a year (if it is a device used infrequently), or a number of failures a month (if used frequently). For this column just define something and stick to it. One of the most common ratings is: 1 – rarely; 3 – occasionally; 5 – regularly. The problem is defining what they mean! The numbers in this column have, again, come from standards (for FMEA it is BS EN 60812 or its IEC equivalent). This is, again, not suitable for industries that range from very small to multinational. Hence Table 7.11(d) should be used for this rating.

The last of the three columns in Table 7.11(b) is the easiest to imagine – "detection." Do not go overboard with numerous ratings as this will make the analysis difficult later. Stick to the 5 point scale but set 1 to be detectable by anyone, with 5 being undetectable (except by someone who has such intimate knowledge – such as you). 3 is in the middle, being detectable but only by someone who is skilled. You may go for finer ratings if you wish, and you can refer to other sources, but I have stuck to these just to make our discussions easier.

However, it is not enough to simply fill in the table; we must do something with it. Figure 7.9 illustrates a typical FMEA process.

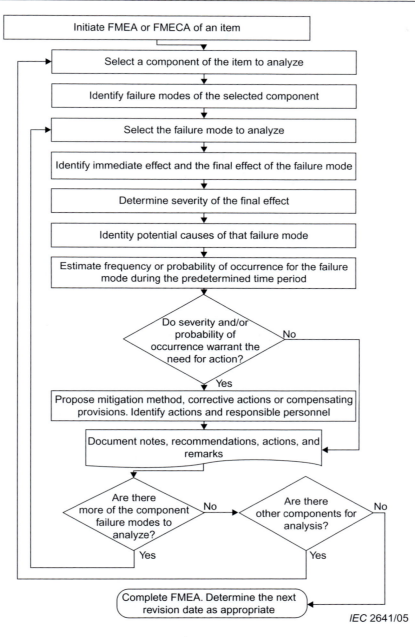

Figure 7.9
Typical FMEA process *(taken from BS EN 60812:2006).*

It is therefore apparent that we need to determine how critical this failure mode is. To do this we determine a Rating Priority Number (RPN) using

$$RPN = Severity \times Occurrence \times Detection \qquad (7.5)$$

We could use a simple number range, for example

0–25	No intervention required
26–50	Intervention required, sign off locally
50+	Intervention required, sign off by senior designer

However this simple rule can hide potential failures. It is therefore better to use a table similar to that illustrated by Table 7.12.

There are three tables, one for each of the three levels of detection. One uses the corresponding table and cross-correlates with the corresponding RPN. Note that the same value of RPN can lead to vastly different outcomes. The numbers in the boxes are not carved in stone; you will need to determine your own tables and only you can do this. There are no guidelines apart from a severity of 5 being bad!

Table 7.12: Example Qualitative Criticality Matrix

(a)

Severity	Detect = 1	Occurrence 1	2	3	4	5
	1	1	2	3	4	5
	2	2	4	6	8	10
	3	3	6	9	12	15
	4	4	8	12	16	20
	5	5	10	15	20	25

(b)

Severity	Detect = 3	Occurrence 1	2	3	4	5
	1	3	6	9	12	15
	2	6	12	18	24	30
	3	9	18	27	36	45
	4	12	24	36	48	60
	5	15	30	45	60	75

(c)

Severity	Detect = 5	Occurrence 1	2	3	4	5
	1	5	10	15	20	25
	2	10	20	30	40	50
	3	15	30	45	60	75
	4	20	40	60	80	100
	5	25	50	75	100	125

The shading illustrates the level of intervention. White means no action is required, the failure mode is controlled; gray means that the failure mode requires some examination to reduce the RPN but it is not mandatory; dark gray means that there must be a design intervention as the failure is not controlled but the changes can be approved locally; black means that the failure is uncontrolled and dangerous design changes must be implemented but approved by a higher level of authority (if the RPN cannot be altered then you really have to decide whether to continue). There are three tables for each level of detection for the simple reason that you could have a failure mode that has an RPN of 25 but is made from $S = 5$, $O = 5$, and $D = 1$; just being able to detect a failure does not make it a safe failure mode.

Some people argue that detection is unnecessary in FMEA analysis – I wholly disagree. Quite often the reason for a high occurrence of a failure is because it cannot be detected. As an example let us look at the current trend of rogue traders bringing down banks. Often they trade using small amounts that are undetectable by the standard trading controls. Their individual severity is small and their occurrence is regular, hence they would have an RPN of 5 and would be deemed safe. How stupid is that? The fact that they are not detectable means that the $S*O*D$ RPN would be 25–50 and at least it would be examined as a problem. If only they had included D in their FMEA we may not be in the financial situation we now find ourselves in during the second decade of the new millennium. Equally, we can use the sinking of the first DeHavilland Comets[1] as an example. Nobody knew about fatigue failure so no one was looking for cracks, hence they literally fell from the sky and at the same time destroyed the UK aircraft industry. Would you be willing to get onto an aircraft that did not have regular inspections for cracks?

You can mitigate a failure mode by including some form of detection that makes the occurrence lower, whereas without the detection the occurrence would be high (this argument is often used to form the basis for the removal of detection from the FMEA chart but it is a flawed argument). Including detection focuses your attention on the end-user: Are they able to detect the potential failure? Have you designed a failure mode safety feature into your device? Detection and occurrence are "soul mates," but you must design them to be so; they need to be introduced to each other not just left to their own devices. And that is why I argue to include detection in an FMEA analysis.

Also, there is another anomaly: that of occurrence. It is possible to design O to be zero. Consider the Titanic: if we designed that ship now and we did an FMEA we would consider the effect of collision with an iceberg. Suppose we suggested that it should only sail in tropical waters…what is the chance of hitting an iceberg in the Caribbean? I would argue zero. It is wholly possible to completely remove a failure mode by being clever and using lateral thinking – but you have to be careful that you don't fall into the trap of thinking that you have designed the unsinkable ship.

---

[1] The DeHavilland Comet was the first commercial jet powered airliner. Its windows were "square" and generated fatigue failure in the fuselage which in turn caused disastrous crashes. Nowadays fatigue is well understood and crack detection is in everything from aircraft to nuclear power stations.

What happens next? If you need to perform an intervention then there will be some form of design change. Either something is modified to be stronger, or some form of detection is included, or even a special instruction is put in the *instructions for use*. Whatever happens there will be a design change that inherently means the FMEA will have changed. This is the reason for the last five columns in an FMEA (as in Table 7.11).

In the first of the five columns one writes a paragraph that describes the change (or the evidence) for the new RPN determined in the last four columns. This is not a word or a phrase – it needs to be complete and succinct.

The second column is the same value of S from the first assessment, simply because the severity of the failure cannot have changed, you only can mitigated against it. Consider a single use item; the obvious danger is mistakenly reusing the device. While you can build in indicators to stop someone from reusing it again, the severity of reuse is still the same.

You should now reevaluate your values of O and D to give new ratings O* and D*. One or the other, or hopefully both, will have been reduced. If you have done your design correctly D* will be 1 and O* will have been reduced. When you calculate your new value of RPN* it should fall in one of the "safe" categories. Note that your design change must reduce O* (we shall see the reason for this when we come to do the mandatory risk assessment later).

CASE STUDY 7.2

Consider the case of a failure mode of a personal insulin pump. Let us assume that the pump has a display to show the time of the last infusion and the amount. Now let us imagine a nasty failure mode. The worst I can think of is that the pump display is showing all is normal, but no insulin is being infused. There is little doubt that this is severe; there is a chance of a patient dying (if they are remote from help, for example when mountaineering, etc.); more likely they will collapse and require urgent medical attention without which the failure could be fatal – hence, from Table 7.11 we have a severity rating of 4.

We now need to determine O. We have little data to go on, hence we resort to looking at the FDA and MHRA websites to identify recalls of similar products. Equally, we can refer to any postmarket surveillance data. We find that over the last year a similar product had four reported failures, hence O = 3.

The detection was assumed to be D = 1 as the patient should feel "wet" as it is leaking under their clothes.

The rating comes out as 12, which is in the "intervention suggested" zone. This means something needs to be done. The infusion line was redesigned with fittings that are tamperproof and robust, and a special band was designed to ensure the "needle" remains intact and in position. Hence the occurrence was reduced to 2. The new table reveals an RPN of 8 which, although not perfect, is acceptable (Table 7.13).

Another, highly acceptable way to analyze RPN is to use a graphical map. If you plot S versus O on a graph you are able to establish contours for various levels of D.

Figure 7.10 illustrates an example RPN plot. As before this is not written in stone; you have to decide your own critical RPN boundaries. This graph does, though, make the decision process

Table 7.13: Example FMEA

Number	Failure	Effect	Cause	Rating				Remedial				
				Severity	Occurrence	Detection	RPN	Action	Severity	Occurrence	Detection	RPN
1	Display shows insulin is being infused but no insulin is entering the body.	Patient can suffer diabetes related episode.	There is a leak in the infusion line.	4	3	1	12					

Severity	Detect = 1	Occurrence				
		1	2	3	4	5
	1	1	2	3	4	5
	2	2	4	6	8	10
	3	3	6	9	12	15
	4	4	8	(12)	16	20
	5	5	10	15	20	25

Number	Failure	Effect	Cause	Rating				Remedial				
				Severity	Occurrence	Detection	RPN	Action	Severity	Occurrence	Detection	RPN
1	Display shows insulin is being infused but no insulin is entering the body.	Patient can suffer diabetes related episode.	There is a leak in the infusion line.	4	3	1	12	Infusion line redesigned to be robust; fittings to be tamperproof.	4	2	1	8

CASE STUDY 7.3

Consider a similar insulin pump but one where the display can become "stuck" due to low battery level. Here the failure is the battery supplies enough power to drive the display, but not enough the power the pump. Hence the system thinks an infusion has been made, whereas nothing has happened. Now the detection is impossible for any but a highly skilled engineer, therefore the value of D has risen to 5. The occurrence of a flat battery is quite commonplace (you should be able to determine the occurrence from the battery rating), so O is 4. The overall RPN is 80, and Table 7.14 shows that this is heavily in the "intervention required" zone and as a consequence must have remedial action. The system was redesigned to incorporate a "buzzer" when the battery is getting low (note not flat). An emergency battery pack was included so that in cases where the charger could not be used the system would still work for 8 hours. This meant that the occurrence has dropped to 2, as it is now rare, and the detection has dropped to 1. The new RPN is 8 and hence acceptable.

Table 7.14: Example FMEA

Number	Failure	Effect	Cause	Rating				Action	Remedial			
				Severity	Occurrence	Detection	RPN		Severity	Occurrence	Detection	RPN
1	Display shows insulin is being infused but no insulin is entering the body.	Patient can suffer diabetes related episode.	A flat battery causes pump to fail while maintaining display activity.	4	4	5	80					

Severity	Detect = 5	Occurrence				
		1	2	3	4	5
	1	5	10	15	20	25
	2	10	20	30	40	50
	3	15	30	45	60	75
	4	20	40	60	(80)	100
	5	25	50	75	100	125

Before

Severity	Detect = 1	Occurrence				
		1	2	3	4	5
	1	1	2	3	4	5
	2	2	4	6	8	10
	3	3	6	9	12	15
	4	4	(8)	12	16	20
	5	5	10	15	20	25

After

Number	Failure	Effect	Cause	Rating				Action	Remedial			
				Severity	Occurrence	Detection	RPN		Severity	Occurrence	Detection	RPN
1	Display shows insulin is being infused but no insulin is entering the body.	Patient can suffer diabetes related episode.	A flat battery causes pump to fail while maintaining display activity.	4	4	5	80	Low battery warning (buzzer) included to activate when 2 hours battery life remaining; emergency battery pack supplied; charger designed to be portable (handbag / pocket size).	4	2	1	8

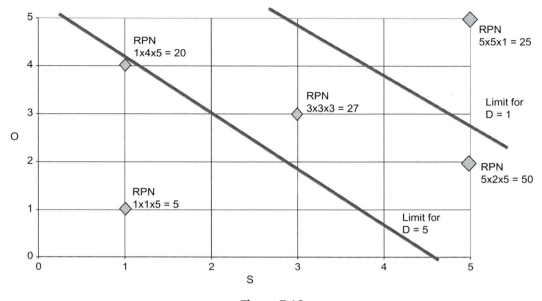

Figure 7.10
RPN presented as a graphical plot.

much easier as you can visualize the S, O, and D and see the effect that D has on the overall RPN. Sometimes it is better to have a graph for each value of D, but it is easy with modern desktop publishing to produce a color-coded contour map with all values of D taken into account.

If you wish to find out more about FMEA then there is a plethora of books on the subject. Also, due to its wide use in industry, the number of websites is enormous. There is FMEA software available, but unless you know what you are doing no amount of software is going to help. The hardest aspect of FMEA is to decide occurrence, detection, and what the value of RPN actually means. The rest is simply imagining the worst.

7.5.1 Fault Tree Analysis

The main issue with FMEA is that you often associate a particular failure mode with a single route cause. In larger systems this need not be the case. More often it is a systemic failure due to a sequence of smaller events. To determine this sequence we use fault tree analysis; as with other important tools there is a standard: IEC 61025 (BSI, 2007). There is another type of fault tree analysis, from Six Sigma – the Ishikawa diagram (or herringbone diagram). Also, there is the classic "5 Whys" we have met earlier. For those of you with devices in the highest classifications a fault tree analysis is a must!

The basis of a fault tree is to determine the sequence of events that leads to this failure. One of the very good reasons for doing this is to design out any potential misuse due to either stupidity or incompetence. Another reason is that this analysis can give us the value for O in an FMEA analysis. This analysis is essential if you use, or intend to use, any software to support your device.

CASE STUDY 7.4

Let us consider a simple computer program which is used to support the use of a device. It is intended, due to its complexity and size, to run the software on a web-based server and the end-user has access to it over the Internet. Without this software the device cannot be used.

One obvious failure mode is that the program stops working. Let us look at the first level of events that may have caused this. The first is that there is an error in the program that causes it to "lock"; the second is that the host server (the Internet provider) fails; the third is that your server fails; and the fourth is that your computer fails (Figure 7.11).

Whatever the failure mode, the overall outcome is that the user will see it as *your fault*. All potential issues are important. Can you see that our piece of software is reliant on other people's equipment – hence if the host computer's wireless connection is faulty do you think the surgeon or clinician is going to hold your software blameless just because it was a hardware problem? No, the procedure is still cancelled – you will still be in the wrong. Hence we use fault tree analysis to imagine what can happen in the overall system.

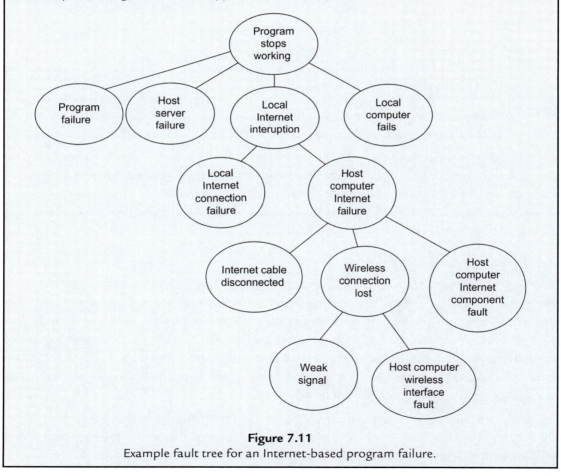

Figure 7.11
Example fault tree for an Internet-based program failure.

Table 7.15: Standard Fault Tree Symbols

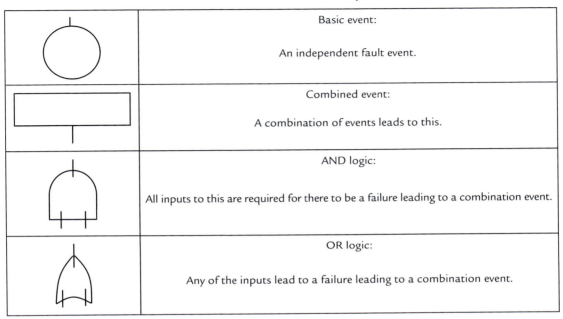

	Basic event: An independent fault event.
	Combined event: A combination of events leads to this.
	AND logic: All inputs to this are required for there to be a failure leading to a combination event.
	OR logic: Any of the inputs lead to a failure leading to a combination event.

However this diagram in Figure 7.11 is too simplistic for a detailed analysis as sometimes things compound together to cause a failure, and sometimes there are basic events that are the root cause. To accommodate for this we have a list of symbols, as in Table 7.15.

If we redraw Figure 7.11 to accommodate these symbols we get the diagram shown in Figure 7.12.

We can use this analysis to develop statistics about the system. This can be horrendously complex and reliability engineers spend a lifetime learning the intricacies. For the purposes of example, let us examine the path laid out in gray. We need to determine the probability of there being a local Internet interruption due to this failure branch. Assuming the individual failures are independent, it is much like the chances of tossing a dice twice and getting two sixes, one after the other.

The chance of getting two sixes are

1st throw 1/6
2nd throw 1/6
$P = 1/6 \times 1/6 = 1/36$

Hence there is a 1/36 chance of getting two sixes.

How many times have you used a computer only to find someone has fiddled with the Internet connection? Let us assume this is a 1/1000 chance. Therefore the probability of this occurring is 1/1000. If this was the only fault then the overall probability of the system failing is 1/1000. But there are another three potential reasons for loss of Internet connection. These are all based on an

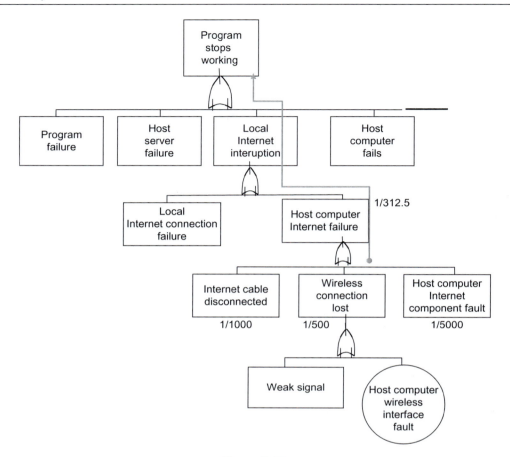

Figure 7.12
Fault tree using standard symbols.

OR logic, in other words any one of these will result in a loss of local Internet connection. If we assume the probability of losing the wireless connection is 1/500, and the probability of the host computer's hardware being faulty is 1/5000, then we can determine the overall probability of the program stopping because of a loss of local Internet connection. If we assume that it is unlikely that all three occur simultaneously, then the probability of one of them happening is given by

Probability of either A or B occurring:

$$P(A + B) = P(A) + P(B) \tag{7.6}$$

If they can happen at the same time then Equation (7.6) becomes

$$P(A + B) = P(A) + P(B) - P(A)P(B) \tag{7.7}$$

CASE STUDY 7.5

If the probabilities of failure in one branch are as described earlier, determine the chances of the system failing due to a host computer Internet failure and hence estimate a value for O (FMEA).

Using Equation (7.6) the probability of host Internet connection failure is

$$P = 1/1000 + 1/500 + 1/5000 = 0.00323 = 1/312.5$$

If this were our only occurrence factor then we can see (from Table 7.11(d)) that this gives a value for O of between 2 and 3. It is, arguably, closer to 2 so we would use this in our FMEA.

More importantly the fault tree helps illustrates all the ways a fault can occur, and it does not let us forget the stupidity of the end-user!

If you wish to know more about fault tree analysis and reliability, I refer you to O'Connor (2002) and Carter (1997), but you need to have a good background in statistics and applied mathematics to follow them. You should also obtain a copy of BS EN 61025:2007 or its equivalent.

CASE STUDY 7.6

A circuit in an OR anesthesia machine was found to fail 1/100 uses. An engineer suggests that using three in parallel would improve reliability. Show that this is true.

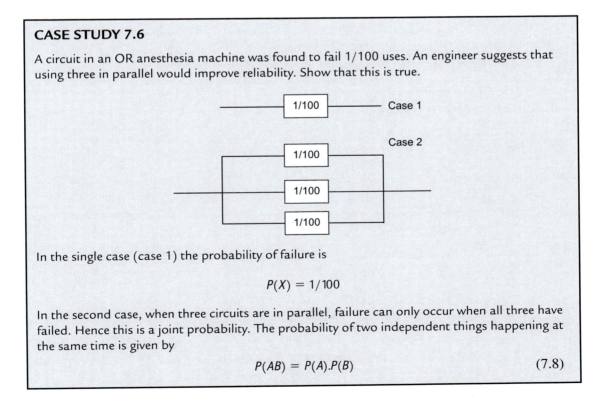

In the single case (case 1) the probability of failure is

$$P(X) = 1/100$$

In the second case, when three circuits are in parallel, failure can only occur when all three have failed. Hence this is a joint probability. The probability of two independent things happening at the same time is given by

$$P(AB) = P(A).P(B) \tag{7.8}$$

Hence the probability of failure is

$$P(X) = 1/100 \times 1/100 \times 1/100 = 1/1,000,000$$

This is called *built-in* redundancy. It is a technique used in the aircraft industry to design robustness into a system.

7.5.1.1 Ishikawa Diagram

Although this is not strictly a fault tree analysis it is another example of a diagrammatic structure that enables a root cause to be identified. The Ishikawa diagram is simple in format but its strength lies in its simplicity. It is commonly called the *herringbone diagram*, because it looks like…a herringbone!

Figure 7.13 illustrates a typical Ishikawa framework (Bicheno & Catherwood, 2005). This format has six main headings: measurements, materials, people, environment, methods, and machines. The idea is to focus on these headings and determine how they may contribute to the failure of your device. Taking each heading in turn:

- Measurements: how data is used and obtained.
- Materials: raw materials used; where they come from and who supplies them.
- People: anyone involved with the making and use of the device and how they interact.
- Machines: dependency of the device on any machines.

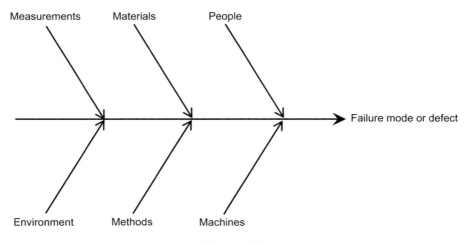

Figure 7.13
Ishikawa diagram.

- Environment: environment the device is in and the effect this has.
- Methods: how device is used and could be misused; instructions that are required.

There are several herringbone models: some with eight arms, some with more, and some with less. It matters not. The main aim of Ishikawa is to get you to think about external factors and how they affect the performance of your device. Once again you can then feed this back into your FMEA to undertake a more detailed analysis.

7.6 D4X

We meet the D4X (Design for X) family of tools when we come to product realization (that is doing the actual "dirty work" of design). However they are also very valuable in design quality too. Let us, for example, examine *Design for Assembly (D4A)*. In the normal context this is taken as designing the device so that it can be assembled quickly and easily on the shop floor. This is fine for the automotive industry but useless for the medical devices industry. How many of our devices are assembled just prior to use? I would suggest the majority. Hence it is a quality issue, as poor D4A will have a detrimental effect on the company's reputation.

Hence, even though the D4X tools are presented in another chapter you should also read them in the context of design quality.

7.7 Six Sigma

I have mentioned Six Sigma (or 6σ) numerous times in this text. There is little doubt that this is the section where it is introduced properly. Without doubt it is the product of the end of the twentieth century (Bicheno and Catherwood, 2005) and is the culmination of disparate quality systems into a coherent strategy. It was first introduced and developed by Motorola with the sole ambition of reducing failures to 6σ, or 3.4 defects per million; an increase of 100× the quality they had previously. The genius of 6σ is in its simplicity; it took all of the quality tools already developed and combined them into a new package. Hence many of the tools used are already well known by designers and engineers. One of the disadvantages of 6σ is that it is heavily policed and to be a 6σ practitioner you must have studied an accredited course and achieved one of their belts (note the link with a martial art!). But the benefit of the widespread adoption is that there is a plethora of textbooks that contain all the information you need to do an adequate job. If you want your company to be 6σ through and through then I'm afraid there is no avenue but to undergo the expensive training programs.

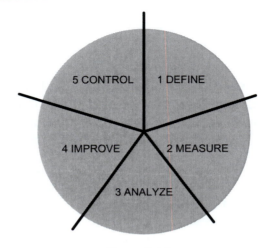

Figure 7.14
The 6σ DMAIC wheel.

There are several building blocks that we can consider in terms of building quality into our design process. The first is DMAIC – Define Measure Analyze Improve Control. I do not intend to replicate a 6σ handbook, but it is safe to say that the terms are pretty obvious. If we are to undertake any quality-based activity, first we should "Define." One could argue that our PDS process has done this. The second activity in the cycle is "Measure," after all if we have defined important issues we should measure our actual outputs to see if we have met them (remember HoQ). The third activity is "Analyze" – just measuring data is not enough – we need to examine the data robustly to determine what has been achieved and what is lacking. The fourth process is "Improve," which follows directly from analysis (DoE and FMEA fit in here). The last is "Control" – it is pointless suggesting all of the former if we never know if they have been done, or if they have been done correctly (remember how much time we spent on procedures).

This is not to say I have given you a Six Sigma design manual, but if you ever attend a Six Sigma course you will understand many of the tools they present to you. Also, it means that you should appreciate that you are well on the way to having a quality product. We shall be meeting some more 6σ tools later.

7.8 Summary

In this chapter we have examined some of the numerous tools used to ensure quality in design. Explicitly we have examined Design of Experiments, House of Quality, FMEA, Ishikawa diagrams, and more. It is important that you put these into practice, as too often people do them "after the event" to try and prove to an auditor that they have a quality design process. This stands out like a sore thumb; it will not fool anyone. More importantly you are actually fooling yourself. The tools are there to help you to develop a robust design that meets the needs of the customer and will withstand normal usage – thus ensuring you will have minimal complaints. An anecdote demonstrates this perfectly.

When the UK automotive industry was at its peak they invited the Japanese to come and see their factories. On the visit the Japanese engineers were taken to the complaints department of one car plant. The UK staff proudly showed the Japanese a vast room with numerous staff taking endless phone calls.

"See how well we manage complaints," said the UK engineer, proudly.

The Japanese engineer looked shocked. "We only have one complaints lady and she has one telephone," he said.

"How can you do that?" asked the UK engineer, "How do you make sure your customer complaints are treated quickly?"

The Japanese engineer replied, "We don't get any."

Enough said I think.

References

Bicheno, J., & Catherwood, P. (2005). *Six sigma and the quality toolbox.* Buckingham: Picsie Books.

British Standards Institute (2006). *Analysis techniques for system reliability – procedure for failure modes and effect analysis.* BS EN 60812:2006.

British Standards Institute (2007). *Fault tree analysis.* BS EN 61025:2007.

British Standards Institute (2009). *Medical devices – application of risk management to medical devices.* BS EN ISO 14971:2009.

Carter, A. D. S. (1997). *Mechanical reliability and design.* Palgrave Macmillan.

Microsoft (2010). *Introduction to optimization with the excel solver tool.* http://office.microsoft.com/en-us/excel-help/introduction-to-optimization-with-the-excel-solver-tool-HA001124595.aspx (cited 16.9.2011).

Montogomery, D. C. (2001). *Design and analysis of experiments.* Chichester: J Wiley and Sons Ltd.

O'Connor, P. (2002). *Practical reliability engineering.* Chichester: J Wiley and Sons Ltd.

## CHAPTER 8

Design Realization/Detailed Design

8.1 Introduction

In many design textbooks you will find this called *embodiment*. I find that term to be confusing to most people. More often than not the books let you decide what to do, as if the design happens by magic. The actual phase we are in is like taking a bunch of ingredients from a cupboard and making a meal; we have asked what everyone wants to eat and now we have to make the dinner.

All of the previous chapters were concerned with taking an individual concept and distilling it down to a single, describable idea. This chapter is about taking that idea and making it "make-able." There is an old saying: "Any fool can make a bolt for $100, it takes an engineer to make it for 1 cent."

This is now the phase of the game that we have entered. We need to take disparate concepts and assemble them as a whole; we need to make sure it can be made; we need to ensure it can be put together; we need to ensure it will withstand the loadings placed upon it; and we need to make sure it works. Hence this chapter aims to present tools and techniques that enable you to "*realize*" your design and make it real.

8.2 The Process to Design Realization

This is hard to quantify as different disciplines and different designs will have their own individual variations. Figure 4.5 in Chapter 4 illustrated a typical detailed design procedure; yours may be different than the one illustrated, but whatever the complexity or simplicity they all follow a general pattern:

1. Macro design realization project plan: identify subprojects and estimate timescales.
2. Assemble design team: identify who you need to help you with the design and get them on board.
3. Micro plan: for each subproject, plan and confirm timescales.
4. Delivery of individual subprojects to timescale.
5. Delivery of overall design realization.

Let us look at these in more detail. But before we do, let us not forget that each one of these little design projects involves a PDS, a creative phase to determine alternatives, and a

Medical Device Design.
DOI: http://dx.doi.org/10.1016/B978-0-12-391942-7.00008-8

selection process to pick the best one and then the analysis starts (just as the overall project did). Do not *hack*!

8.2.1 Macro Project Plan

While we would have a project plan for the whole project it is worth revisiting this to ensure that this stage works smoothly. We may have allocated 6 months to this phase; we need to make this work so we produce an overall macroscale project plan in which we estimate the timescales for all of the subprojects. It cannot be stressed enough that most detailed designs live and die based on whether the project has been managed properly or not.

8.2.2 Assemble Design Team

Hopefully we will have already assembled a team at the start of the project. However, if this is not the case, we really do need one now. One designer cannot be master of all. We will need support from analytical engineers, from packaging designers, from materials experts, from a whole range of disciplines. Now is the time, if you have not already done so, to select those team members. We shall look at this process in more detail later.

8.2.3 Micro Plan

Each subproject will have its own timescale. It is important that you lock this down at the earliest opportunity or you will find your overall project dragging. This is not only costly for the company, but it could cost you your job.

8.2.4 Delivery of Subprojects

By now you will have realized that the essence here is the development of a good PDS. You will have established what the subproject needs to do and what it needs to meet. Hence delivery also means acceptance.

All of the planning and hard work you put in earlier should now come to fruition. If you have carried out your design planning correctly the subproject undertaken by this individual should work smoothly and the transition from "brain to paper" should be seamless – and rapid.

8.2.5 Delivery of Overall Design

Now the real sweat starts to be produced. Does the design go together as expected? If you have followed all of the tools I have demonstrated earlier, and use the tools in this chapter, we should have success, first time – every time.

8.2.6 How?

This is the real question. How do we do this? How does it happen? Hopefully the next few sections will help you to start a process. There are certain things you *have to do* and certain things you will do through experience. In the end, as with riding a horse, the more you do it the better you get; and if you do fall off – get straight back on!

8.3 Assemble Your Detailed Design Team

As I have already stated, hopefully you are following a holistic model and you have thought about this much earlier in the project. However, even in the most holistic of models it is worth confirming your design team at this stage. The first thing to decide is who is the lead designer? They are the team coach; they are the team's manager; they are the film's director. They assemble the best team for the job.

Assuming you are the lead designer, here are some questions you should ask *yourself* in order to decide your team membership:

- *Will I be using any subcontractors to supply elements of the design?*

The most common people here are companies who supply sterilization trays, people who supply sterile packaging solutions, and people who supply transportation packaging. But do not forget the people who produce your instructions for use, the people who produce your labels, and the people who will actually make the item.

- *Who is going to be making the device?*

Will you be using subcontractors? Do they need to have certain certificates (e.g., ISO 9001, or ISO 13485)?

- *Do you need technical support?*

Are you able to do the design calculations? Are you able to perform any necessary experiments? Do you need material selection advice?

You need to think of your design team as a sports team. If you were coaching a baseball team you would not have a team that could only hit a ball, with no pitchers. If you were coaching a soccer team you would not have 11 goalkeepers on the pitch. So in your team you need a balance of skills. Each project will have its own demands so the main aim of the "lead designer" is to decide what skill set is required and then match this against people. A good way of doing this is to produce a skills requirements map, as illustrated by Table 8.1.

You cannot foresee all potential requirements but you should be able to envisage major requirements. It is also important to not only identify who has the real skill in that area, i.e., the team's expert, but also those who have had experience of working in that area (say from a previous project). You will be amazed how this latter level of skill is beneficial; if only not to reinvent the wheel.

Table 8.1: Typical Design Team Skills Map

Skills / Body	Relevant Certification	Use CAD	Undertake Design Calculations	Perform DoE	Sterile Packaging Expert	Materials Selection
Employee 1	n/a	X	O		O	
Employee 2	CEng		X	X		
Company 1	ISO 13485				X	
Research Body 1	ISO 9001					X

Key: X = expert; O = has experience in this area

8.3.1 DHF Considerations for the "Lead Designer"

You will have appreciated, from previous chapters, that all regulatory bodies want a design history file (DHF). Hence one of your main roles as lead designer is to ensure that the DHF is populated and updated. This means that you need to ensure that you receive documentary evidence from all of the participants, and that documentation is in the form you require.

Hence another consideration for the lead designer is contracts. You cannot run this level of project on "word of mouth" instruction; all must be documented. Hence the two external bodies in Table 8.1 need to have been selected on merit (documented); the PDS of their aspect of the project has to be agreed and signed off; and your contract with them must ensure that they provide you with the information you require for your DHF.

A more subtle aspect of dealing with external companies is security. You should always have a non-disclosure agreement with all of your subcontractors. If the project is highly secret you should ensure that they know it is so and that all is kept secret – do not rely on thinking they know.

The final thing the lead designer needs to do is to organize regular design meetings to review progress and agree on any design changes. These need to be regular, and with today's Internet availability they can be done using video-conferencing. Again, all of this needs documenting, agendas need to be set, and action plans need to be produced and monitored. Everything needs to go into the DHF.

Even if you are doing the design all by yourself all of the above requires consideration as the DHF must exist; you cannot escape the need for a populated DHF no matter how small your company. While it is difficult to have meetings with yourself you *will* have meetings with subcontractors at some time; you will make design decisions. They *all* need documenting in the DHF.

8.3.2 Phases of a Team

Because teams are so important to many industries and social networks, there has been much research conducted to look at how they operate and how to make them operate better. This is out of the scope of this book but this does not mean that as lead designer you should ignore it; on the contrary, you should read as many books about teamwork as possible. You do rely on them, after all.

Tuckman[1] developed the concept of the four phases of a team (Tuckman, 1965) (Tuckman & Jensen, 1977):

Phase 1: FORMING – In the first stage of team development, team members want to know "What is expected of me? How do I fit in? What are we supposed to do? What are the rules?" Anxiety can follow initial excitement. No one feels secure enough to "act as themselves" so there is not much open conflict. You will need to set operating guidelines or ground rules. You also need to spot prima donnas.

Phase 2: STORMING – Initial enthusiasm gives way to frustration and anger. The team struggles to find ways of working together and everyone seems awkward. What appear to be resistance, wrangling, hostile subgroups, and jealousies arise. Ground rules start to be broken. A hard phase to get through, but get through you must.

Phase 3: NORMING – Gradually, the team settles and enters the "norming" phase. Team members start to find independence and standard ways to routine things; power plays and grandstanding become less evident. Team members may hold back good ideas for fear of introducing conflict. Help the team by giving them responsibility and authority.

Phase 4: PERFORMING – In the fourth and final phase, the team goes about its business with smooth self-confidence. Team members disagree constructively; they take measured risks, and have real ownership for their work. They also appreciate what the other members of the team are doing and rather than seeing this as being in conflict to what they're doing, they see it as a benefit. The team can experience "storming" periods at any time – when under unusual pressure, for example. The team can also return to the "forming" phase (especially if new members join).

Your role as lead designer is to get to the *performing* stage as fast as possible, and then keep your team there. I have found that by giving team members ownership of the project, straight away, they respond quickly. Hence your first team meeting is critical. How you run this meeting is down to you but a good "icebreaker" helps; being focused on the problem at hand also helps. To use another soccer analogy, if the team manager says, "Okay lads we're going to win this one; go out and sock it to them," what response do you expect compared with "Okay lads this game's not important – win, draw, lose no problem"? As they say in China "A one thousand mile walk starts with one step"; make sure the step is in the right direction because it's a long way back! There is a plethora of textbooks and general reading books on team building/teamwork – I leave the choice of reading to you.

---

[1] See, for example, http://en.wikipedia.org/wiki/Tuckman's_stages_of_group_development

8.3.3 Design Meetings/Design Reviews

Whatever you do, do not forget that all of your activities must be auditable. If you recall Figure 4.2, in Chapter 4, you will remember that the year must contain a series of design reviews. To ensure that your team produces zero nonconformities you, as lead designer, must ensure that they are following procedures. Hence one good part of any design meeting is to have a standing item related to quality management. As the lead designer you are able to perform small audits on your team to make sure that procedures are followed. As a part of a bigger quality management process you need to be prepared to be audited too.

You need to ensure that your design meetings have a set agenda. Have standing items that tie in with the company's quality management process. A seamless transition is always beneficial.

Apart from the obvious QM role, the design meetings are essential to maintain communication between the whole of your design workforce. Do not worry about too much communication; it is when someone goes quiet that you need to worry! The lead designer must take on this "paternal" role for the project because, as described in Chapter 1, we are delivering the baby. The obvious aim of the design meeting is to ensure the project is on track and that everyone is keeping to timescales; the obvious connection with QM is to make sure no one is "cutting corners."

8.4 Design Calculations

It is hard to envisage any design that does not contain at least one design calculation. The calculation may be very simple or it may be very complex. The complexity matters not, it is the process and the documentation that matters.

All design calculations should follow these steps:

- State what you trying to solve.
- State all assumptions made, giving reasons and references.
- State the equation(s) used, giving reasons and references.
- Perform the calculations, writing down every step.
- State the answer to the calculations.
- Sign off by the person "calculating."
- If necessary sign off by a qualified person (e.g., chartered/licensed engineer).

As with most things of this nature, we have seen that the best method is to have an established pro forma and make sure everyone uses it. Figure 8.1 illustrates a typical pro forma for design calculations; feel free to make your own.

Obviously the rows can be expanded to suit. Also the sections could easily cross pages so it is important to have the "header" on all pages. Note that there is a section for "conclusions"; you must analyze your calculations and make suggestions from them, if only to say something as simple as "it must be a minimum of 10 mm diameter."

Co. Name	Product:		
	Part:		
Calculations performed by:	Date:		Sheet: X of Y

Description:	
Assumptions:	
Model / Equations:	
Calculations:	
Conclusions:	
Signature: Name: Date:	Checked: (if appropriate) Name: Date:

Figure 8.1
Typical design calculations pro forma.

It is important that the "calculator" signs off all design calculations; they may need signing off by someone in authority (as described earlier) so make sure a provision has been made for both.

Note that many calculations are now performed using computer models. Most industrial systems have the capability to produce a report similar to that of Figure 8.1. If the one you are using does not then you must produce one yourself. It is a common failing of people using computer models to only give a final answer; this is not enough. Programs such as MathCad® produce a report as you build the model. Finite element packages, such as ANSYS, also have a reporting function. The trouble with all of these is that you must include the "assumptions" and "conclusions," so make sure you do.

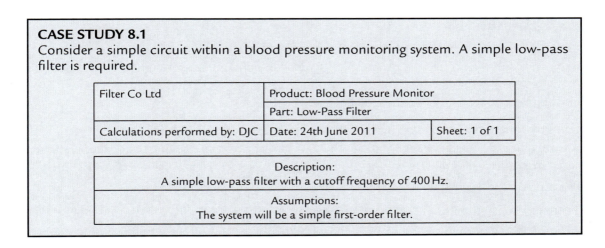

CASE STUDY 8.1
Consider a simple circuit within a blood pressure monitoring system. A simple low-pass filter is required.

Filter Co Ltd	Product: Blood Pressure Monitor	
	Part: Low-Pass Filter	
Calculations performed by: DJC	Date: 24th June 2011	Sheet: 1 of 1

Description: A simple low-pass filter with a cutoff frequency of 400 Hz.
Assumptions: The system will be a simple first-order filter.

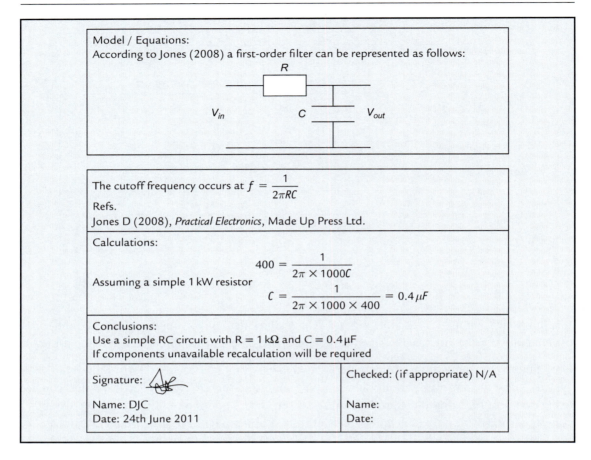

Model / Equations:
According to Jones (2008) a first-order filter can be represented as follows:

The cutoff frequency occurs at $f = \dfrac{1}{2\pi RC}$

Refs.
Jones D (2008), *Practical Electronics*, Made Up Press Ltd.

Calculations:

$$400 = \frac{1}{2\pi \times 1000C}$$

Assuming a simple 1 kW resistor

$$C = \frac{1}{2\pi \times 1000 \times 400} = 0.4\,\mu F$$

Conclusions:
Use a simple RC circuit with R = 1 kΩ and C = 0.4 µF
If components unavailable recalculation will be required

Signature:	Checked: (if appropriate) N/A
Name: DJC	Name:
Date: 24th June 2011	Date:

You should be referring to classic textbooks for your "human" specific data and models (see Section 5.3.1.4.1 – Biomechanics).

8.4.1 Computer-Aided Analysis

In the previous section I mentioned some typical computer-aided analysis packages. Whereas only a few short years ago they were solely available to multinationals and universities, most (bar a few) are within the reach of most medium enterprises and certainly, if you have a university close by, will be in the reach of everyone.

I will try to introduce them to you shortly, but there is one important concept you need to fully appreciate. This concept is called GIGO – "garbage in, garbage out." Very simply, these software packages all rely on your assumptions being correct. Hence the use of a structured pro forma (as described previously) is essential.

There is also another trap which is commonly called *painting by numbers*. Because these analyses tend to produce pictures with lots of colors, they can fool the uninitiated into making the wrong deduction (most commonly red is bad because red is danger); or even worse into thinking lots of colors must mean the answer is right. You can be badly misguided by "keyboard jockeys" (KBJs): those who own an analysis package but do not know how to use it, *properly*. Unfortunately, due to the prevalence of available software and the misguided notion that they are simply tools, there is also a prevalence of KBJs.

If you are using any computer-aided analysis packages make sure that whoever is producing the model and analyzing the results knows what they are doing. Today there are too many KBJs in the marketplace. If you are going to buy this service treat it as if you were going for a major operation, i.e., make sure they are a qualified surgeon not that they have watched every episode of *Scrubs*.

8.4.2 Computer-Aided Analysis Disciplines

While Table 8.2 attempts to convey the variety of analysis packages available, it is impossible to illustrate all of them. However, what it does do is open your eyes to the wealth of computer-assisted calculations. A word of caution: freeware and apps should be avoided. As a medical devices manufacturer you should always use software that has been audited in some form. You have no idea where an app or a piece of freeware has been written, so you have no idea if it works or it doesn't. However, and MATLAB has started this, it is perfectly feasible for the systems listed in Table 8.2 to be available for smartphones and tablets in the very near future.

If you are going to use one of these software items do not forget to put the decision and reasons why in your calculation report.

To give you an idea of what can be achieved using modern CAA packages, Figure 8.2 has been devised to illustrate an FEA analysis of a hypothetical implant. It is loaded at one end and is held at the other by locking screws. The storyboard of figures illustrates how the analysis takes the solid model derived from CAD (see later in this chapter) and produces an FEA model from which a variety of solutions can be determined. In this case total displacement (bending) and Von-Mises stresses have been illustrated. Using an accepted FEA program is as valid as any hand calculation, but it is only one part of a rigorous analysis.

There is such a variety of computer-aided analysis packages available that virtually any design can be analytically verified wholly by simulation. Indeed it is said that most modern aircraft are validated in that fashion. However, as with aircraft, no one would consider being the first passenger on the maiden flight if it had not flown somewhere already – hence it is unlikely, in the near future, that CAA totally replaces physical verification. But it should be stressed that using CAA saves time and hence cuts costs.

Figure 8.3 illustrates another typical CAA package: CFD. In this case a fluid is flowing into a converging–diverging section of a pipe. Once again the storyboard illustrates how the model is built. In this case velocity of flow and pressure profile are illustrated as flow lines.

Table 8.2: Some Typical Computer-Aided Analysis Packages

Mathematical Analysis (Including Optimization)	
MAPLE	Designed for mathematicians. Has the ability to solve equations numerically, but also to solve them as equations. Can produce a report for the DHF.
MATHCAD	Like MAPLE but designed specifically for engineers; hence incorporates toolboxes suited to engineering calculations. Can perform statistical analysis for hypothesis testing.
MATLAB	Designed for matrix manipulation. Large selection of toolboxes for control, simulation, and signal processing. A great tool for data analysis.
EXCEL	Spreadsheet: great for DoE analyses. Also incorporates some statistical analysis for hypothesis testing.
SCILAB	A freeware version of MATLAB but with much reduced functionality.

Data Analysis/Data Logging	
LABVIEW	Primarily a data logging program written by National Instruments. It is a world leader in this discipline. However its control simulation and its ability to act as a data analysis tool make it very useful.
Computational Fluid Dynamics (CFD)	CFD has grown in popularity over the past two decades. CFD can be used to analyze fluid flow over bodies, in bodies, or through bodies. Has clear use in pulmonary, cardiovascular, and thoracic medicine.
CHAM - PHEONICS	A world class CFD package with the ability to model a variety of fluidic systems.
FLOTRAN	Similar to CHAM but comes as part of the ANSYS suite of software.
CFX	An easier to use analysis system built into ANSYS Workbench suite.
Finite Element Analysis (FEA)	FEA has grown from its inception in the middle of the twentieth century. It is used in everything from structural analysis, heat transfer, electromagnetics, and noise.
ANSYS	A stand-alone world class versatile FEA package particularly well suited to structural analysis and heat transfer – but has many other functions. Incorporates optimization.
COSMOS	Part of the SolidWorks suite. FEA and fluid flow are incorporated with similar tools as ANSYS.
MSC NASTRAN	Another world class FEA solver but comes with a 3D dynamic modeling system that allows dynamic forces to be modeled and integrated.
ALGOR	Another world class FEA system.
ABAQUS	Especially good for nonlinear materials, such as rubbers.

Electrical/Electronic	
PSpice	Enables electrical circuits to be developed and analyzed.

8.5 Materials Selection

There is little doubt that selection of an appropriate material for your device, or part of your device, will give you some sleepless nights. Essentially there is one thing you need to think of straight away: What has been used before? This section will, I hope, help you to select suitable materials for your designs using a structured approach – and not just rely on a "gut feeling" (which is wholly the wrong approach).

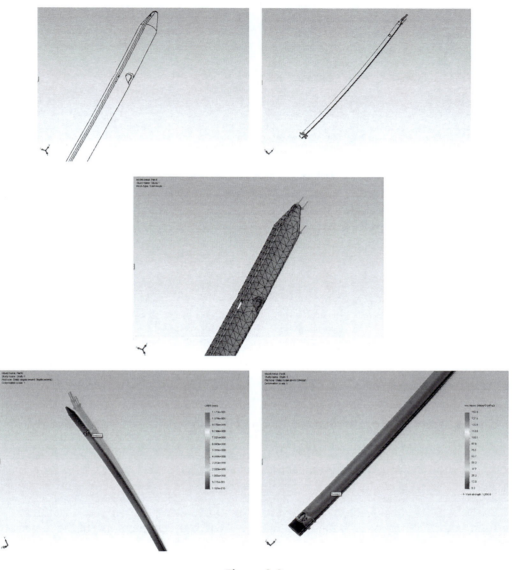

Figure 8.2
Example finite element analysis of an implantable device under axial loading: (a) solid
model; (b) constraints and loading applied; (c) solid model discretized into a mesh of "finite
elements"; (d) results for displacement (note original shape is also shown); (e) predicted
Von-Mises stresses enable failure by yield to be detected. (Performed using COSMOS in
SolidWorks)

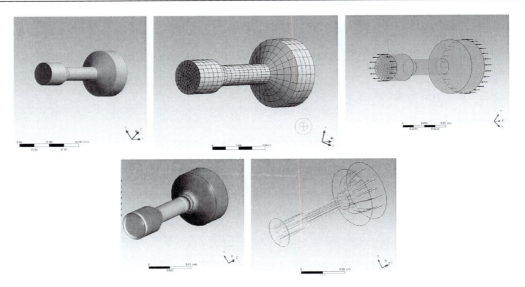

Figure 8.3
Example CFD of fluid flowing in a convergent–divergent pipe: (a) solid model; (b) CFD mesh;
(c) input/output constraints; (d) predicted pressure profile; (e) predicted velocity streamlines.
(Performed using CFX in ANSYS)

8.5.1 Formalizing the Selection Process

Selecting a material is not a simple task. It is difficult, complex, and requires justification. Hence it is advisable to use a similar pro forma to that presented for design calculations.

Figure 8.4 illustrates a typical pro forma. If completed fully, it is able to go into the DHF directly. As with other such pro forma, additional sheets are of course possible.

8.5.2 PDS

Hopefully your first port of call, the PDS, will give you some pointers, if not the material itself. If not it will certainly have given you the design constraints to go into your materials selection pro forma.

8.5.3 Precedent

Do you remember way back in Chapter 2 I introduced the idea that finding the classification for your device can be made easier by finding a precedent? The same applies to materials. If your device is similar to that of someone else, and that device is in common use, then does it not make sense to start with that material? Of course, in some circumstances the company will keep material data confidential, but in most cases it is easy to find. Be sure to check, however, that there are no outstanding recalls!

Co. Name	Product:		
	Part:		
Materials selection performed by:	Date:		Sheet: X of Y

Description:	
Assumptions:	
Information sources:	
Conclusions:	
Signature: Name: Date:	Checked: (if appropriate) Name: Date:

Figure 8.4
Typical materials selection pro forma.

Do not forget that if you have been in medical devices for some time your own experience counts, as do your own precedents.

8.5.4 Research

Now we are going back to Chapter 5, "Developing Your Product Design Specification." Here, if you remember, I introduced the concept of the *data cloud* (Figure 5.4). I also introduced the concept of the mini-PDS for particular items. Your research to generate this mini-PDS will almost certainly provide information that gives you pointers to materials that can be used. It is almost certain that your end-users will have a good idea of commonly used materials, as will your manufacturing chain.

However one area of research, often untapped, is scientific journals. Many companies have their products tested versus competitors by a university research group who then publish the findings. These papers often contain material specifications! Equally, there is a wealth of clinical research papers that look at the performance of these devices; the same applies. The third type of paper is one which looks at issues related to certain devices (and often their materials); again, the same applies.

If you conduct a thorough research project you will, almost certainly, find clear pointers to materials that can be used. To start, one of the best books for materials selection is Ashby (2004).

If you are researching a new material that has never been used in this type of device before, then some desk and laboratory research is essential. Tying up with a university not only brings independence, it can also bring match funding. Indeed, many universities offer materials expertise as a service to industry.

8.5.5 *Regulatory Bodies*

Another good source of information is the regulatory bodies themselves. In many cases you can pull down guidelines that provide pointers to good practice, materials that have been passed for use in certain areas, materials that are not allowed, and pointers to standards that may be applicable (although these should have been found in the PDS). An alternative to this is to ask your supplier if the material has been approved for use by one of the regulatory bodies – take care though because food use and medical use are not necessarily the same! The FDA website is, once again, a wonderful resource, if you know your precedents. A simple 510(k) search (as demonstrated previously) reveals a plethora of information related to your proposed device, as illustrated by the search sequence in Figure 8.5.

8.5.6 *Standards*

Almost without question is the fact that there will be numerous standards listing what materials are to be used in certain situations. In fact, it makes sense to make this your first

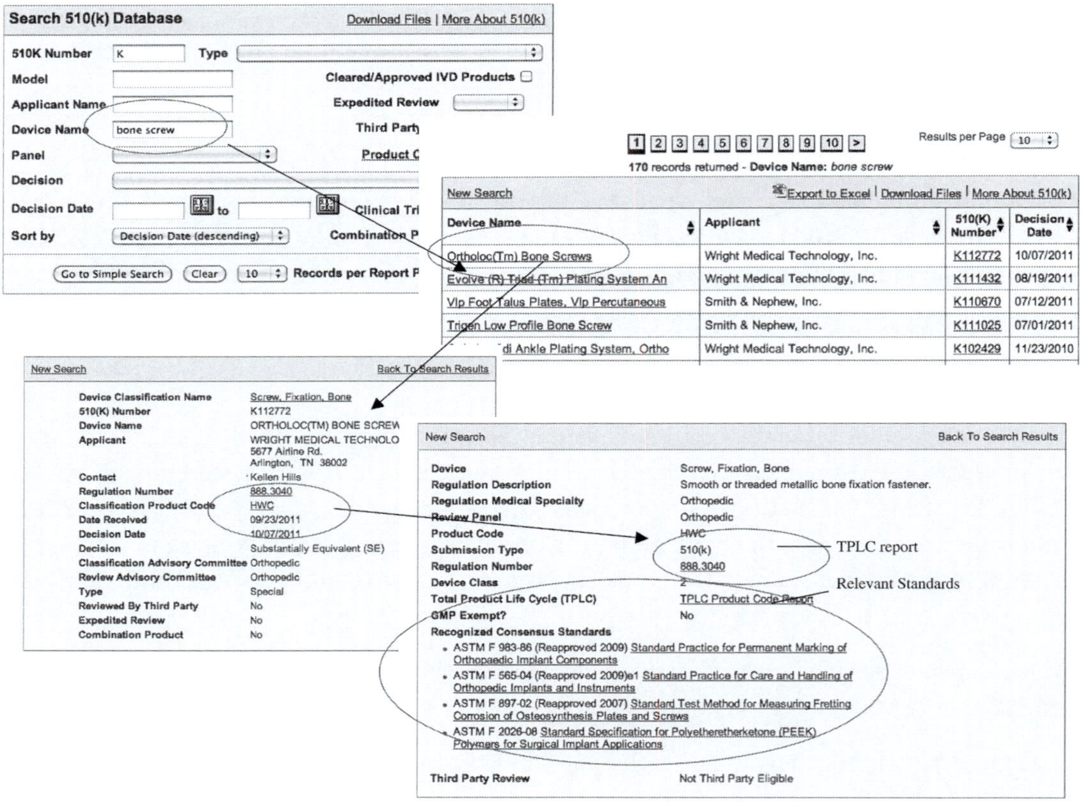

Figure 8.5
Using the 510(k) search engine to determine applicable standards for a particular device.

port of call (as your PDS should have alluded to). Take care because standards in the EU are different to those in the USA and you may need to check compatibility. The FDA website has a consensus database you can search: http://www.accessdata.fda.gov/scripts/cdrh/cfdocs/cfStandards/search.cfm.

All standards bodies have a web presence that is open to all. You are free to search at will. For example, Figure 8.6 is a screenshot from the British Standards Institute of the results of a standards search related to materials for suture wires.

The search has yielded ISO 10334, an international standard that must therefore have an ASTM equivalent. This standard will probably list those materials you can choose from, and those alone.

If there is a standard like this then your life becomes easy! There are a number of standard standards. One of them is ISO 7153-1:2011 which is the standard for "*Surgical Instruments. Metallic Materials. Stainless Steel*": this is obviously a must for any surgical instrument maker. Nearly every discipline has standards like this one, so find yours and get hold of it.

8.5.7 Materials Search Engines

There are two I would like to draw your attention to. The first is MatWeb (http://www.matweb.com). This is an online materials search engine that is vast. It will not be able to direct you to the exact material, but it will enable you to examine materials in fine detail, obtain material properties, and pull down typical uses and suppliers. More importantly,

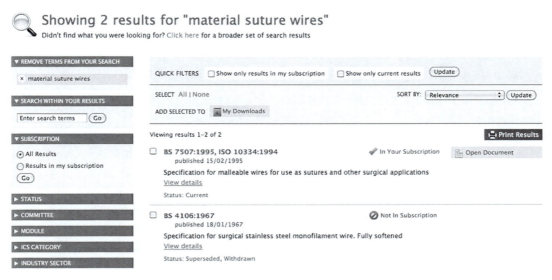

Figure 8.6
Illustration of a standards search for "material suture wires."

Table 8.3: Example Material Search Table

Property	Range of Values
Density	$<1800\,\text{kg}/\text{m}^3$
Yield strength	$>500\,\text{MN}/\text{m}^2$
Water absorption	Low
Gamma irradiation	No effect
Max operating temperature	$40\,^{\circ}\text{C}$
Min operating Temperature	$-20\,^{\circ}\text{C}$

it will state if it is used in the medical industry and sometimes even what for! For a fee you can also link the materials database direct to ANSYS and SolidWorks (described earlier).

The second is not a search engine but a materials selection package developed by Cambridge University (UK). CES (Cambridge Engineering Selector) enables you to enter design criteria and then potential materials appear. More importantly, it enables you to put in the design criteria in such a manner that not only do potential materials appear, but it is possible to select the optimum material using merit indices (as described earlier). It is beyond the scope of this text to teach you CES, but you will find most engineering departments will have access to its full potential.

When using search engines you need to be structured. You can do random searches but this normally ends with everyone using stainless steel as everyone simply ends up there. Instead be very strategic. Develop some search criteria as illustrated in Table 8.3. You will find that you will have to undergo some rigorous design calculations first.

As with previous advice, a good PDS will have provided most of this information. However each specific search will have more detail associated with it (such as yield strength) that cannot be known at the PDS stage; the detail will only be made clear after some calculations have been performed.

Once you perform your search you will have a range of materials to pick from. You need to order these with some form of merit index. You are free to use the weighted selection criteria demonstrated in previous chapters. Equally, Ashby (2004) proposes the use of merit indices. If you come from an engineering background the merit indices are easy to understand; if you do not then use the weighted selection criteria table. One thing you can all do is plot a graph. Let us, for example, suppose that you really need a material that has low water absorption properties but high yield strength. If you plot a graph of one property versus the other, and then use points to represent the material, you may get a graph similar to that in Figure 8.7. Clearly the materials you really want lie in the bottom right-hand corner. Figure 8.8 illustrates a typical Ashby merit indices graph – in this case Young's modulus versus density. Some common medical device materials have been indicated. The shaded areas demonstrate where typical material families reside on this graph. If you plot a line (such as the ones indicated)

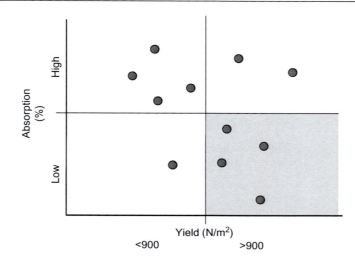

Figure 8.7
Example materials merit graph.

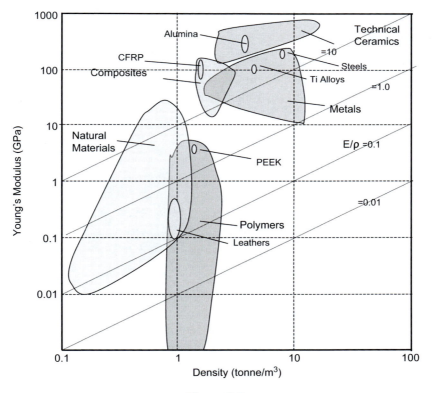

Figure 8.8
Example of an "Ashby type" merits indices graph indicating some common medical devices materials. *(Source: Ashby, 2004)*

that describes the properties you want, then those materials lying on, or near, this line are the ones that suit your specification.

The lines illustrated are various ratios of E/r: as the ratio increases, stiffness compared with density increases. Thus those lying above a line are stiff and light, and those below are flexible and heavy. For example, if your PDS stated that the ratio of stiffness to weight should be between 1 and 10 then only PEEK, from the materials shown in Figure 8.8, falls into this category.

You should refer to Ashby (2004) for more details of this selection methodology. If you use materials data sources correctly (such as MatWeb) you can build graphs of this nature for yourself using actual material data, rather than general data.

8.5.8 Advisory Bodies

Virtually every material has an advisory body of some form or other. These bodies are, almost always, not-for-profit organizations and are there to encourage industry to make use of their materials. Therefore if you have an idea that you want to use a ceramic, then hunt down your ceramics advisory body and simply ask. Remember, good designers are not afraid to ask!

8.5.9 Consultancies

If you feel you are out of your depth, then the advisory bodies mentioned previously will have lists of consultants they are able to recommend. Of course, you will have to pay, but on the other hand if the material selection is wrong it is their fault and the burden of blame shifts to them.

8.5.10 Animal Products

We cannot escape this. All regulatory bodies now ask if your device has any animal products (includes human and nonhuman sources) associated with it. This used to be confined to those that had some derivative tissue, or an agent that had been developed from an animal source. However the fear of prions and their resistance to normal sterilization methods means that all animal derivatives come into play. Unfortunately, animal products are used heavily in manufacturing, especially the manufacturing of plastic products.

If you are going to make your life easy then avoid animal-based products; this means you must provide assurance that your whole manufacturing chain also avoids the use of animal products. You must ask the question, "Do you use animal-based products, such as lubricants, in your manufacturing process?" You must get the answer "No," or you have lots of work to do to prove its safety.

If, however, you *must* use an animal-based product then you should follow the relevant guidelines laid down by the FDA and MDD. For example, an FDA guideline can be found at http://www.fda.gov/MedicalDevices/DeviceRegulationandGuidance/GuidanceDocuments/ucm073810.htm.

8.5.11 Biocompatibility

I do not intend to teach you every term related to biocompatibility. We will meet the important issues in a moment. However, just because you use stainless steel does not mean you have considered biocompatibility. If you followed the guidance given in Section 8.5.1 then biocompatibility should not be an issue (all of the hard work has been done for you). Even if your material is deemed "safe" an eagle eye must be kept pointed at recall notices, the news channels, and any other sources of "product recall" information – just in case the material concerned is also used in your device! However, if you intend to use a material that has not been used before then that is a different ball game!

You must not underestimate the importance placed on biocompatibility. You must be able to provide evidence that the materials you use comply with the essential requirements for a medical device. That is why the materials selection pro forma is so important to your materials selection records.

8.5.11.1 Scope
Biocompatibility comes into play when a device is communicating with tissue. Do not assume this is only by direct contact; it can also be indirect (it could be by emission or vapor). If your device communicates with tissue then its biocompatibility must have been tested at some time. Biocompatibility testing is an expensive, time-consuming business and that is why following Section 8.5.1 and using the efforts of others helps!

8.5.11.2 Definition(s) and Standards
Initially it was thought that only *inert* materials were biocompatible, but the definition has changed due to the increase in using substances that can be implanted into the body that are not inert (e.g., resorbable implants). There was much debate but the current definition of biocompatibility states that biocompatibility is

> *The ability of a material to perform with an appropriate host response in a specific application.*
> **(Williams, 1999)**

This is known as the Williams[2] definition and was adopted by the European Society of biomaterials. It is brief and to the point. Basically it states that if you have a material communicating with a host, it should do what it is supposed to do without harming the host – as in this further definition that expands on the former (not surprisingly, written by Williams):

> *Refers to the ability of a biomaterial to perform its desired function with respect to a medical therapy, without eliciting any undesirable local or systemic effects in the recipient or beneficiary of that therapy, but generating the most appropriate beneficial cellular or tissue response in that specific situation, and optimizing the clinically relevant performance of that therapy.*
> **(Williams, 2003)**

---

[2]Prof. Williams of Liverpool University, UK. His book *The Williams Dictionary of Biomaterials* is probably one you should have on your shelf.

Table 8.4: A Selection of ISO Biocompatibility Standards and FDA Recognized Consensus Standards

Standard	Title
BS EN ISO 10993 family	Biological evaluation of medical devices.
BS EN ISO 10993-1:2009	Biological evaluation of medical devices. Evaluation and testing within a risk management process.
FDA Recognition Number 2-156	Biological evaluation of medical devices. Evaluation and testing within a risk management process (biocompatibility).
BS EN ISO 7405:2008	Evaluation of biocompatibility of medical devices used in dentistry.
FDA Recognition Number 4-179	ISO 7405 – Evaluation of biocompatibility of medical devices used in dentistry.
BS EN ISO 11979-5:2006	Ophthalmic implants. Intraocular lenses. Biocompatibility.
FDA Recognition Number 10-48	Ophthalmic implants. Intraocular lenses. Biocompatibility.

Therefore if you are intending to use a material that communicates with a body then it has to be proven to behave in the way it is supposed to behave, and that it does not cause any deleterious effects. Note these effects can be local or systemic: this means the material could simply produce a skin rash at the point of application; or it could bring on an asthma attack even though the lungs have not been touched. Also, the effects may present themselves many years later. This is a serious issue for the introduction of novel materials: How do you identify age-related effects?

It should be no surprise that there are numerous guidelines and standards associated with this topic. A good rule of thumb is to use what has been used before unless there is a very good reason not to do so.

Table 8.4 is very brief. ISO 10993 has 16 parts, all having slightly different nuances on biocompatibility. They are revised all of the time as new evidence of effects comes to light. Even now Part 3 (genotoxicity, carcinogenicity, and reproductive toxicity) is under revision and is open for public comment. Clearly you should have a copy of Part 1 at hand.

You will also note that there are specific tests for particular disciplines. If, for example, you were in dentistry you must also meet the relevant standards for that discipline too. Make sure that you do this search for your own discipline and for the state or country in which you intend to sell, be that the USA, the EU, or wherever.

8.6 Computer-Aided Design

The rise of the personal computer (be that PC or Mac) has been relentless. But a few years ago companies had darkened rooms containing two or three very expensive workstations running bespoke software. Nowadays personal desktop computers are extremely powerful, and the software is extremely user friendly and affordable. Hence, the rise of computer-aided design has been equally relentless. There are a few front-runners in the CAD field, which are:

AutoCAD®
CATIA®
ProEngineer®
SolidWorks®

All have their pros and cons, and I do not dare to demonstrate any preference over one or the other. But the one thing they all have is the ability to produce realistic 3D visualizations of the design and transfer electronic information over the Internet. Why are these two items so important? Let us go back 20 years. A draughtsman would produce a two-dimensional engineering drawing that had to be copied to be transferred; you had to know how to read the drawing to understand it; and you had to have extremely good 3D spatial awareness to imagine the 3D shape drawn out as a 2D plan drawing. Not really conducive to collaborative work!

Let us now compare the modern equivalent. The CAD draughtsman produces a 3D object perfectly representing the component; this is transferred to 2D drawings for manufacture or to a 3D model for rapid-prototyping (see later); an electronic email version can be sent to all partners who *do not* need the software to see the design and do not need to be able to read drawings. I think you can imagine the power I am trying to demonstrate.

Nowadays we are also able to have online discussions around the design; you can have someone produce an electronic circuit in Japan and then integrate this with your model (in front of you online) and check if it fits! Yes, this is all possible. If you are not using CAD in your design work, then you really are in the dark ages.

It is difficult to convey the power modern CAD systems have. It is only when you have seen a system in action that you begin to realize just what you can do. Also, when you think that a good desktop PC may cost between $2000 and $3000, and the software may cost $6000, it's not really anything to balk at. Modern CAD systems really do save time and are worth the investment (Figure 8.9).

Before we go any further we need to understand two concepts: solid models and surface models. A solid model is just that: if you draw a cube you have a solid cube. Hence solid modelers produce solids. A surface model system still produces a solid component, but it is hollow. Hence, when you draw a cube you do not get a solid cube,

| 2D sketch visualization | Solid model | Colored solid model | Photo-ready |

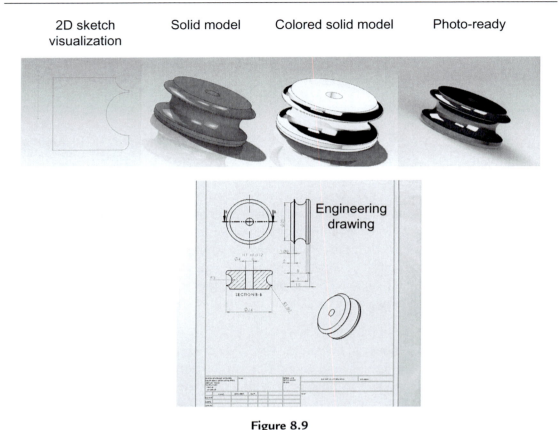

Figure 8.9
Progression of CAD (for a simple component) from 2D sketch through 3D photo-ready visualization to production-ready engineering drawing.

you get a cube with a void inside. The two have come about because of the way CAD has developed; both have strengths and weakness, but most people will be using solid models at the start.

8.6.1 *Cloud Computing*

A modern trend is to use *cloud* computing. For the designer using a CAD system this is invaluable. In the past keeping track of the most up-to-date design was a nightmare. Cloud servers have made this a thing of the past. A central server holds the information securely and it synchronizes when anything changes. Therefore the cloud server always holds the most up-to-date information. Even more importantly, it looks after the backup too! If you have ever lost any work due to forgetting to back up you will know how valuable this is.

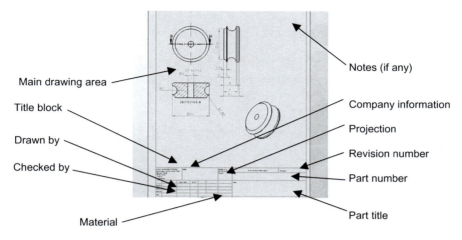

Notes (if any)

Main drawing area

Company information

Title block

Projection

Drawn by

Revision number

Checked by

Part number

Material

Part title

Figure 8.10
A typical engineering drawing containing relevant documentary evidence.

Table 8.5: International Standards for Technical/Engineering Drawings

Standard Number	Scope of Standard
BS 8888:2008	Technical product specification – Specification
ISO 1101:2004	Technical drawings – Geometrical tolerancing – Tolerance of form, orientation, location and run-out – Generalities, definitions, symbols, indications on drawings.
ISO 1660: 1996	Technical drawings – Dimensioning and tolerancing of profiles.
ISO 2692:2006	Technical drawings – Geometrical tolerancing – Maximum material principle.
ISO 5458:1999	Geometrical product specification (GPS) – Geometrical tolerancing – Positional tolerancing.
ISO 5459:1981	Technical drawings – Geometrical tolerancing – Datums and datum systems for geometrical tolerances.
ISO 8015	The principle of independency and the envelope requirement.
ISO 10578:1992	Technical drawings – Tolerancing of orientation and location – Projected tolerance zone.
ISO 2768-2:1989	General tolerances.

8.6.2 Document and Revision Management

Most modern CAD systems also come with built-in revision management. Once again, this is often in conjunction with a secure server (some of them use the cloud servers above). Most programs force you to keep revision histories up to date. Figure 8.10 illustrates the key components of an engineering drawing, but you should refer to the relevant standard(s); some are given in Table 8.5 for guidance.

8.6.3 Collaboration

As I stated earlier, collaboration is easy once you have gone digital. Most modern CAD systems come with interfaces, such as EDrawings®, that enable you to have discussions about a design without the need for everyone to have an expensive copy of your software (EDrawings is free). However you should be aware of two types of collaboration:

Asynchronous: This is commonly someone looking at a document, drawing, or file, then commenting and sending it back or sending it to a group. This is the most obvious type of collaboration if the partners are on opposite sides of the world. This is the most common form of collaboration with CAD.

Synchronous: This is where partners look at each other, at the device, and comment at the same time. This is clearly easy if all are in the same (or similar) time zone. Some CAD systems come with this capability but most are asynchronous. Once again the Internet comes to our rescue and software such as Skype enables us to have collaborative meetings with relative ease.

One thing to be sure of, however, is data transmission. While programs such as EDrawings help with communication they do not help with trials. If your manufacturer wants to try out the design, this type of file does not help. Normally you would have to send your design in another format. The most common type is an IGES file. This file contains a solid model of your design and most CAD packages can import them with ease. Equally, most CAD packages also import everyone else's file, but this is often more troublesome than using an IGES file.

Don't forget that this is still a meeting. As such you need documentary evidence for your DHF!

8.6.4 Reverse Engineering

I think you can see that the introduction of CAD is really starting to do wonders. Modern CAD systems have the ability to import a digital "point cloud." This point cloud is a collection of x-y-z coordinates that describe the outer profile of a 3D object. Some typical sources of such data are coordinate measuring machines, laser scanners, and CAT scan data. If you are designing things to fit a human body then knowing the actual dimensions of that body part is really useful. This point cloud data brings that shape into life; it is easily converted into a surface model, which can then be converted to a solid. If you think this is all far-fetched consider that maxillofacial surgeons are using this technique to develop implants that are unique to the patient. Figures 8.11–8.13 illustrate typical devices to generate a point cloud. Figure 8.14 demonstrates the transition from a point cloud to a solid model. Most CAD packages do this automatically, to lesser or greater effect.

8.6.5 Engineering Drawings

At the very end, something has to be made. Therefore no matter how pretty your component looks in CAD, an engineering drawing will be required for your DHF. Your CAD system must have the ability to produce engineering drawings (to the relevant standard) with relative ease (see Figure 8.10). Remember these are controlled documents so they need revision numbers and if they change, just as with any document, a record of what has

Figure 8.11
Free-standing laser scanner and turntable. *(Courtesy Staffordshire University, UK)*

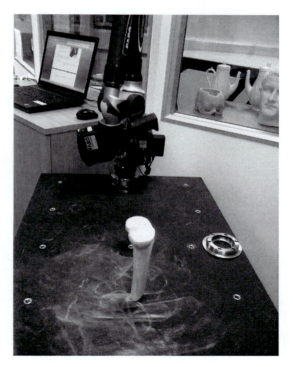

Figure 8.12
Handheld "robotic arm" laser scanner. *(Courtesy Staffordshire University, UK)*

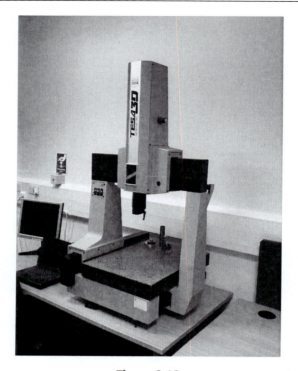

Figure 8.13
Commercial coordinate measuring machine. *(Courtesy Staffordshire University, UK)*

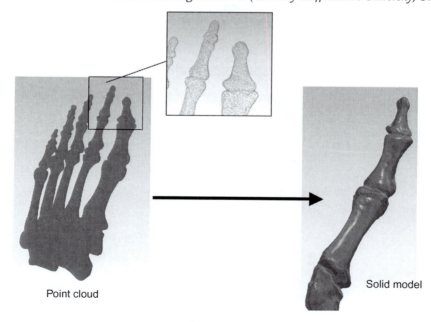

Point cloud

Solid model

Figure 8.14
Typical "point cloud" obtained from a laser scanner converted to 3D "solid."

changed must be made. Also you need to make sure that everyone is using the correct versions, probably more so than in any other part of your processes, otherwise all will fall apart. Hence make sure you control the release of engineering drawings and part specifications with the utmost care.

8.6.6 Part Numbering

Some countries, such as Germany, demand that all parts of a medical device (if they are removable) should be individually numbered. *Logical part numbering* is a bone of contention with everyone. You need a sensible part numbering system that does not just suit your stores but also suits your packers and the end-user. Once devised most CAD packages will allow you to use the part numbers as your filing system. But it is so easy to go down an illogical path. You will find part numbers that are totally intelligible for the stores, but the end-user has no idea what is going on. My advice is to think of the end-user, or the person putting it together. If the part numbering is logical it should help assembly. If the part numbering is simply sequential (in the order drawn) it is not much better than useless. Choose wisely.

One part numbering system I have seen that seems fairly simple for the designer and the end-user (but not all) is to use outlining as with textbook chapter titles. First you need to allocate an overall identifier for the main device, say, X1; thereafter, just as with outlining, the number grows in length. Hence:

> X1000 is the part number for the overall device;
> X1100 would be the part number for the 1st subassembly;
> X1200 would be the part number for the 2nd subassembly, and so on.
> If X1400 also had sub-subassemblies then these would be X1410, X1420, and so on (as illustrated in Figure 8.15).

Notice that the last number in the first four digits is a zero – this indicates it is an assembly. You need to decide how many digits there are in this sequence to allow you to accommodate all of your subassemblies.

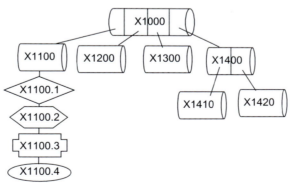

Figure 8.15
Illustration of the concept of part numbering.

Let us now assume that sub-assembly X1100 comprises four parts. These individual parts would then be numbered X1100.1, X1100.2, X1100.3, and last of all X1100.4. If the main body also had a single component (such as a covering sheet) this would then be X1000.1. Simple isn't it? All part numbering systems need to be simple, and work for the designer and the end-user, not just for the warehouse staff.

The benefit of this system is that it is easy for all healthcare staff to know which bits go together. Imagine if they were numbered X1234256, X156719, X17823076, and X145629a: Would you be able to tell that they were from the same assembly? It also helps your packing staff to understand what goes with what, and hence minimizes packing errors.

This numbering system may not be to all tastes. Also some hospital purchasing systems cannot cope with the usual – ;. / symbols used as divisors, so make sure that no number is identical even with the symbols removed. Also watch out for mixed text and numbers – the number 1 can so easily be confused with the letter l, etc.

You need to enforce a logical part numbering system from the outset – do not leave it for your stores to sort out at a later date. Renumbering parts at a later date is really difficult, and really time-consuming.

8.6.7 Tolerances

Once again, most CAD packages allow you to build tolerances into your components. However, when you are producing your models at the start take care as you will set a fixed value. Hence if you have a shaft that is 10 mm diameter +0.1 mm and −0.0 mm, what do you use for the initial solid? Do you start with 9.9 mm (bottom), 9.95 mm (middle), or 10 mm (top)? Some would argue that if you want the tolerances to work, always draw to middle values. Others say stick to the original. Whatever you decide, stick to one method. I have to admit that for some critical tolerances I use the "draw to middle" concept as it seems to make the shop floor happy as they like to work to the "main dimension."

Once again, and this is where CAD really helps, all forms of tolerancing and respective options should be built-in.

8.6.8 Sign Off

Do not forget that the drawings of your components and assemblies are controlled documents. They all need revision numbers (stated earlier), but they will all need to be signed off as valid. This need not be done on paper; nowadays CAD packages allow for electronic sign off. I am old fashioned – I like to have a paper copy in a file, but for larger companies this would be a mountain of paper (imagine the number of drawings for a Boeing 747) and hence electronic data sign off is acceptable.

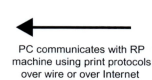

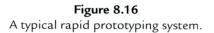

PC communicates with RP
machine using print protocols
over wire or over Internet

RP produces 3D solid in "hours"   CAD produces 3D model (usually STL)

Figure 8.16
A typical rapid prototyping system.

8.6.9 Rapid Prototyping (RP)

Once again, the rapid growth of the PC has made this technology viable. It is just as easy, nowadays, to print a solid object as it is to print text on a page. Even the machines look similar!

The basic principle (as illustrated in Figure 8.16) is the same as an inkjet printer, except that in most cases instead of printing ink you are laying down layers of ABS. After numerous layers a 3D solid emerges that is close to your design. The more expensive the RP machine the more accurate the model. Do not think that only plastics are used; you can RP in ceramics and metals. I have seen RP products in stainless steel and in titanium. I have even seen results of tests that suggest that they are as strong as solid models. I have even seen RP solids in hydroxyapatite. Now we really are opening the doors to bespoke human engineering. Figures 8.17–8.18 illustrate some typical commercial RP machines.

8.6.10 3D Visualization

Some of you may have seen 3D movies at the cinema. This technology is heavily used in the design environment. Many modern CAD systems enable the user to produce 3D simulations of the object concerned, just as in the cinema and as illustrated in Figure 8.19. While the designer may think this nonsense, for the nondesigner and the end-user this is highly valuable to provide early feedback. The same issues apply as to 3D cinemas; you need specialist 3D projection systems. But the rise of 3D television has made this more attractive to design houses. At this moment in time not all CAD software has this ability and you may need to export items to specialist visualization software. The time will come when this form of visualization becomes commonplace, and even the designers will probably be working in virtual-3D as the norm.

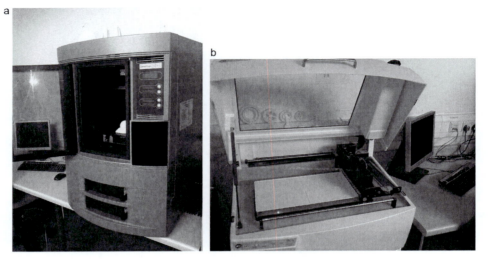

Figure 8.17
Commercial rapid prototyping machines: (a) ABS; (b) "ceramic." *(Courtesy Staffordshire University, UK)*

Figure 8.18
Commercial 2D laser profiler. *(Courtesy Staffordshire University, UK)*

8.7 D4X

I have spoken much about DFX principles. Hopefully you have been waiting with anticipation for them to arrive. Your waiting is over! DFX simply means Design for X, where X is a discipline. It is hard to know where it started but most likely it started with Design for Manufacture. We now have a list that covers Design for Assembly, Disassembly, Manufacture, Sterilization…and so on. In this section we will look at disciplines that I feel

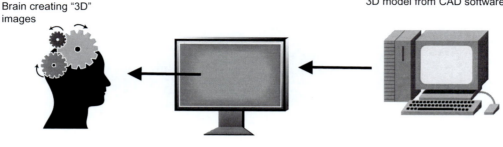

Brain creating "3D" images

3D model from CAD software

Image from 3D TV screen or projection

Figure 8.19
Representation of a 3D visualization system.

Table 8.6: 6σ Seven Wastes

Waste	Description
1. Waste of overproduction	What does overproduction mean to you?
2. Waste of waiting	Time costs money.
3. Waste of transporting	Are you moving things around too often?
4. Waste of inappropriate processing	Only if necessary?
5. Waste of unnecessary inventory	Stock on the shelf means less cash in the bank.
6. Waste of unnecessary motion	Ergonomics and positioning – effort.
7. Waste of defects	Defects cost money – not only replacement but also repair contingency funds.
"New" Waste	
7(a) Waste of making the wrong product	Is your product valued?
7(b) Waste of untapped human potential	Are you using all skills available?
7(c) Waste of inappropriate systems	Are you using things just because you can?
7(d) Waste energy	Are you wasting energy?

are pertinent to medical devices. Before we do so, let me introduce you to another useful 6σ tool: *the seven wastes*. If you use these as guides DFX becomes easier as the whole point of DFX is to avoid waste. The seven wastes are a lovely tool to help you avoid "overdesigning" your device. Table 8.6 introduces the seven wastes, and their newer siblings.

8.7.1 Design for Manufacture (DFM)

Very simply, your device has to be made. There may only be one; in which case DFM is not important. However I am hoping that your device will be made in bulk. If this is the case DFM is very important. The best way to consider DFM is to examine common failings that led to its inception. Consider the component in Figure 8.20. The designer produced a lovely CAD

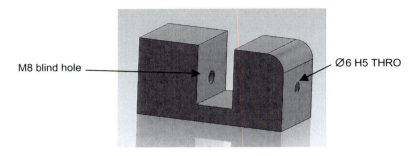

Figure 8.20
Ill-conceived design with no thought for manufacturing.

model that works "on paper" but completely forgot how it is going to be made (not just made but made efficiently and cheaply). How is the M8 hole to be machined? There is a piece of metalwork in the way – and the 6 mm hole is not big enough to allow a tap to be introduced.

Hence the designer has not thought about how it is going to be made!

Now consider a component with several holes, all of different diameters. Are these diameters really necessary? Can we reduce the number of diameters? "Why?" I hear you ask. Because the operator (or the machine) has to change the drill for each hole, and this wastes time and hence wastes money. Also the manufacturer has to stock all of the drills, many times over. All of this adds to unnecessary cost that will drive up the cost of your device. Do not forget, it is not only the material used in your actual component, you also have to add how much scrap your produce: you pay for the scrap! As shown in Figure 8.21 this can be very costly!

Once again, consider the component illustrated in Figure 8.20. Does the hole have to be such a tight tolerance? Can its tolerance be lowered so that a more simple process can be used? A general procedure to process these questions is illustrated by Figure 8.22.

How is the machine operator expected to hold the device? You may think this silly, but if you are having your components coated then where they are held becomes important as that point will have no coating!

You will notice that this is where communication with your potential manufacturer comes into play. Failure to perform DFM is the main reason for most design failures at the prototype stage; it is where you get the embarrassing phone call "and how do you expect us to make this?" If you bring manufacturing expertise into play as soon as possible then DFM becomes very easy indeed.

Some questions to consider, in addition to those in Table 8.7, are:

- How is it likely to be made (what process)?
- Are there too many variations of design features?

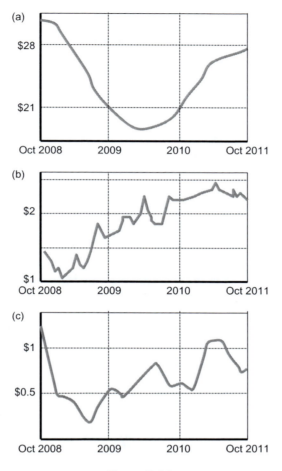

Figure 8.21
Typical variations in material costs $/kg: (a) titanium; (b) aluminum alloy; (c) cold rolled steel.
(Source www.metalprices.com)

- How will the component/item be held?
- Can every feature be made?

As with most things modern, some CAD packages come with DFM options built in (see Figure 8.23). All of this has been built on the back of the work of Boothroyd and Dewhurst,[3] who were awarded the National Medal for Technology for their efforts (shown to save U.S. industry billions of dollars). For more detail I refer you to Boothroyd et al. (2010).

---

[3] Boothroyd and Dewhurst started their own DFMA company based on computerization of the DFM and DFA process: http://www.dfma.com/

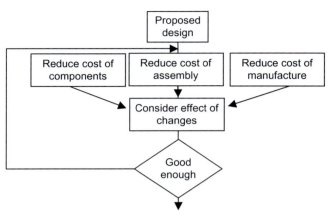

Figure 8.22
General DFM(A) process.

Table 8.7: 6σ Seven Wastes Applied to DFM

Waste	Description/Question
1. Waste of overproduction	• Can you reduce off-cuts? • How many components can be made from one standard size item (sheet, bar, etc.)? • How many do you need to make?
2. Waste of waiting	• How long does it take to make? • Time to delivery? • Will you need to wait excessively for specialist materials/services?
3. Waste of transporting	• Will it need multiple subcontractors to complete? • Are the subcontractors disparately placed?
4. Waste of inappropriate processing	• No. of holes? • No. of variations? • Silly dimensions or tolerances? • Do you need new tooling or can you reuse existing tooling?
5. Waste of unnecessary inventory	• Too many components? • Too many variations in stock items? • Too many variations of similar components? • Do you need new jigs and fixtures or can existing ones be reused?
6. Waste of unnecessary motion	• "Fiddly" surface profile? • No real point of holding during manufacture? • Have you specified a datum?
7. Waste of defects	• Too complex? • Designed in manufacturing issues? • Ill-conceived tolerances?
"New" Waste	
7(b) Waste of untapped human potential	• Have you spoken to the people who will make it? • Have you sought advice from those who may understand the manufacturing process?

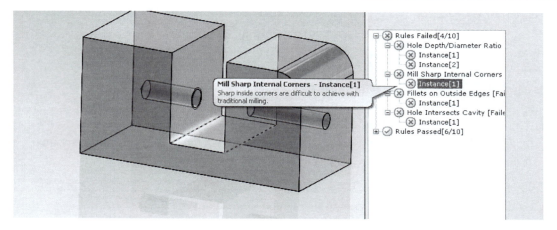

Figure 8.23
Sample DFMA output showing internal edges are sharp. (Performed using SolidWorks)

Another aspect you may not have appreciated is that there may be some manufacturing required by the end-user. Consider implants that require some form of cement or adhesive. This adhesive may require some mixing and a way of placing it correctly; is this not manufacturing? Yes it is – hence the DFM stretches to the end-user too. Have you made the manufacturing process easy? Is it intelligible? If you perform this analysis correctly it will make your life easier when we come to do the risk analysis later.

As a further hint, do not forget the lessons of Pareto[4]: 80% of your manufacturing costs will be related to 20% of your components. You need to identify the 20%.

8.7.2 Design for Assembly (DFA)

The obvious starting point for multinationals putting together numerous components is to look at the assembly process. Hence it is arguable that this could have come first; however I suspect DFM was the first to be done. Although this has a massive impact on large batch manufacturing (such as printed circuit boards and automobiles) it also has a massive impact on medical devices. We shall explore…

One of the main concepts in DFA is components. Quite often the designer will use a standard catalog and produce a plethora of different components. Too many variations of a component cause two sets of issues.

The first is on the factory floor: too many components cause problems with purchasing and storage, and are a frequent cause of mistakes (especially if they look the same). It

---

[4] Vilfredo Pareto observed, in 1906, that 80% of the land in Italy was owned by 20% of the population; hence we now know this ratio as Pareto.

makes rapid assembly (especially by automated systems) virtually impossible. This is not much of an issue if your device has two components, but if it contains an electronic printed circuit board then it will have lots. The obvious thing to do is to reduce inventory, that is, make as many components as alike as possible, for example by using the same bolt, the same nut, and the same pneumatic fitting throughout. It is obvious when pointed out, but is so often missed. One should also look at numbers of items. You may have a lid held onto a box with 20 bolts; are they all required? How many do you really need? It seems small, but each bolt has a cost and putting in the bolt may cost even more than the bolt itself!

Another aspect is how your device goes together. Is it easy to assemble or is it difficult? Are jigs and fixtures required? If so, design them. Can you design out the need for jig and fixtures? If our device's method of operation depends on good assembly, then this cannot be ignored.

The second aspect is assembly at the point of use. Many of your devices will require some assembly by the end-user; some may need complete assembly from a box of bits. Do not forget these people are unskilled in manufacturing – have you considered their assembly demands in your design? You must imagine yourself in their place; do not just assume it is easy for them. This is also a really good point to get the end-users in and see if your assembly protocol is logical.

Table 8.8: 6σ Seven Wastes Applied to DFA

Waste	Description/Question
1. Waste of overproduction	• Too many components? • Too many assembly tasks?
2. Waste of waiting	• Too complex assembly that requires significant training? • Specialist operations or tooling required?
3. Waste of transporting	• Too many subcontractors required? • Too many site visits required?
4. Waste of inappropriate processing	• Too many operations? • Silly, ill-conceived joining methods?
5. Waste of unnecessary inventory	• Too many components? • Too many variations of similar components?
6. Waste of unnecessary motion	• "Fiddly" assembly methodology? • Ill-conceived assembly protocol? • Too many variations of similar looking components? • No thought of end-user assembly?
7. Waste of defects	• Too complex? • No thought of end-user issues?
"New" Waste	
7(c) Waste of inappropriate systems	• Use of confusing technology? • Inappropriate MMI?

For example, if you have components that look the same but are different, how are they supposed to know? The classic failure here is knobs that look the same but which have slightly different threads.

Do they need special tools to assemble your device? Are there critical components that can be misassembled?

As you can imagine, if you have assembly at point of use then your risk analysis has to be robust. Hence thinking about this in a DFA analysis actually makes *your* life a lot easier.

Some questions to consider when performing a DFA analysis are:

- How many different components do we have? Are they all necessary?
- How many fixings do we have? Are they all necessary?
- Do we have any components that look the same and hence can be confused?
- Are any jigs and fixtures required?
- Is the assembly logical and easy to perform?
- Is assembly dependent on highly skilled technicians?

Table 8.9: 6σ Seven Wastes Applied to DFDA

Waste	Description/Question
1. Waste of overproduction 2. Waste of waiting	• Poor joining methodology leads to excessive scrap? • How long does it take to disassemble? • Will you need to wait excessively for specialist materials/ services? • Is significant training required?
3. Waste of transporting	• Will it need multiple subcontractors to complete? • Are numerous site visits required?
4. Waste of inappropriate processing	• No. of variations? • Ill-conceived joining leading to inappropriate unjoining methods? • Have you used adhesive where nonpermanent fixings would suffice (or vice versa)?
5. Waste of unnecessary inventory	• Too many items required for replacement post-disassembly?
6. Waste of unnecessary motion	• Ill-conceived protocol? • No thought of person doing disassembly? • Fiddly process?
7. Waste of defects	• Too complex? • Designed in issues? • Disassembly leads to scrap?
"New" Waste	
7(b) Waste of untapped human potential	• Have you spoken to the people who will disassemble it? • Have you sought advice from those who may understand the usual processes in situ?

8.7.3 Design for Disassembly (DFDA)

This is a relatively new concept that has come into force since recycling of materials has become an important feature of household items and cars. However the lessons are important to medical device designers. You must remember that some of your devices will break and they will need repairing; how easy is it to take them apart to do the repair?

Some of your devices will be repeatedly assembled and disassembled by the end-user. We considered assembling them in the previous section; but after use how easy is it to take them apart? Will the disassembly process damage the components – even if not intentionally?

Where do the components go to when disassembled? Are they placed somewhere logical so that when reassembly is required the components are easily found? Are they easily detectable (e.g., different colors)?

Many designers forget this simple but logical part of the design's life.

8.7.4 Design for Sterilization (DFS)

This is a special part of DFX just for the medical devices industry. If your device needs to be sterilized, which it most likely will be, how does the sterilization process affect your design?

First you need to consider which of the three common methods of sterilization will be used: steam, gamma irradiation, or ethyl oxide. Clearly each method has its own deleterious effect on specific components, but your design must withstand these.

Are you sure your device can be sterilized? If it is reusable does it actually fit into a standard sterilization system? There are trays of all sizes that can be made, but which do your customers want? Do they have a preferred footprint? The tray will need marking so that the theater staff can identify any missing components. You may wish to have a purpose-built tray with supports so that your device does not rattle around in an empty box. All the above needs considering beforehand. If your device is reusable it will, with most probability, be repeatedly steam sterilized. How will repeated exposure to steam at 130°C affect your materials or components? Will some components need to be replaced after each sterilization?

One aspect of sterilization most people forget is cleaning. All devices are cleaned before being sterilized. Can your device be cleaned? Blind holes are not recommended, nor are very long, small bore holes (for obvious reasons). Anything trapped during the cleaning process will be sterilized. It will be rubbish (called *bio-burden*), albeit sterile rubbish; or at least it should be – remember that if your bio-burden levels increase significantly then the sterilization process is at risk of failure and this puts your patients at risk. The last thing anyone wants to see is your device emitting "gunk" – sterile or not. This is not

Table 8.10: 6σ Seven Wastes Applied to DFS

Waste	Description/Question
1. Waste of overproduction	• How many washes are required for acceptable bio-burden levels? • What level of sterilization is required to ensure sterility?
2. Waste of waiting	• Are blind holes, etc. included that will require specific tests to be undertaken?
3. Waste of transporting	• What facilities do the end-users normally have on site? • Have we designed in a system that requires specialist centers to sterilize and hence involves transportation?
4. Waste of inappropriate processing	• Are the sterilization protocols standard?
5. Waste of unnecessary inventory	• Do the end-users need special tools to clean and sterilize your device?
6. Waste of unnecessary motion	• Packing devices into standard trays? • Packing and unpacking by central cleaning/sterilization services (CSD)?
7. Waste of defects	• Can CSD lose components? • Is the device sterilizable by "normal" CSD units?
"New" Waste	
7(b) Waste of untapped human potential	• Have you sought advice from those who may understand the usual processes in situ?

restricted to holes; bio-burdens can get trapped in machining marks, poorly assembled joints, and gaps.

Many cleaning/washing regimes are very harsh (Ph13/14 is common). They can destroy aluminum components at a whim. Can your device survive the cleaning/washing process? This part of DFS is very easy – talk to the people who actually do the cleaning/washing.

Questions you should ask are:

• How will my device be sterilized? What affect does this have on my device?
• Will it be reusable?
• Will washing/cleaning affect my device or its components?
• Can my device be washed properly? Are there any blind holes? Are there any long cannula? Is there any chance of bio-burden collecting?
• Are there cracks/joints that may cause bio-burden issues?

As with previous sections, spending time on this aspect will help with the risk analysis required later.

8.7.5 Design for Environment/Sustainability (DFE)

If you look in any design handbook before about 2000 you will find a cursory glance at this topic, if anything at all. It is a truism that before we became aware of our impact on the

Table 8.11: Energy Utilization for the Production of Some Common Materials

Material	Primary Production (MJ/kg)	Secondary – Recycled (MJ/kg)	CO_2 (kg/kg)
Aluminum	130–260	47–160	8.48–24
Ferrous	13.4–29.2	9–14	1.32–2.3
Paper based	12–41	13–21	0.0009–0.0027

(Extracted from Grimes et al., 2008)

Table 8.12: 6σ Seven Wastes Applied to DFE (or Efficiency)

Waste	Description/Question
1. Waste of overproduction	• Does your device produce unnecessary waste by-products – can they be reduced? • Does your packaging use more material than it needs to? • Is your design optimal?
2. Waste of waiting	• Are your end-users "doing" nothing for long periods while using your device? • Does your device stand idle for long periods but still use energy (standby mode)?
3. Waste of transporting	• Are you flying components all over the world? • Are your components being processed efficiently and in similar locations?
4. Waste of inappropriate processing	• Have you designed your components to be made efficiently? • Are they optimized? • Do your end-users have numerous processes to go through to reach the device's ultimate conclusion?
5. Waste of unnecessary inventory	• Are you using too many parts? • Can some parts be made common? • Can you go modular? • Are all of your packaging items the same or all different?
6. Waste of unnecessary motion	• Are your components designed in order to be made efficiently? • Do your end-users have to perform numerous difficult functions to use your device?
7. Waste of defects	• Is your device evaluated to ensure potential failures have been designed out? • Is your manufacturing QM system robust enough to ensure 6σ type outcomes?
"New" Waste	
7(a) Waste of making the wrong product	• Are you listening to your customers and end-users? • Do you really know what they want?
7(b) Waste of untapped human potential	• Are you including your whole data cloud to influence your design? • Are you using the experiences of others to help your design? • Are you using the "human body" to provide an energy source?
7(c) Waste of inappropriate systems 7(d) Waste energy	• Are you throwing a computer at a device for no real reason? • Are you including energy-hungry subsystems for no real reason? • Have you really investigated the whole device, how it is going to be used and it's actual life (whilst in use) to ensure that it is efficient?

environment engineers and designers were guilty of wastefulness. However we have become the saviors in this respect as only we can design the solutions to the problems. One way in which this manifests itself is in "carbon footprints" and "zero to landfill" policies. In the medical devices environment (at the time of writing) this is difficult as there is little chance of recycling, say, a used hip. But many of us will be designing medical devices for the mass consumer market and have no such excuse. ISO 14044 is the standard associated with life cycle analysis (LCA). It is a standard you should have at hand if you intend to go down the LCA path. A cursory glance at how much energy is used to make and recycle common materials (Table 8.11) is "eye-opening."

Most countries have an environmental policy in place, and in most this exhibits itself in a statute of some form. To cover the statutes in detail is beyond the scope of this text but you should realize that as a trading company you will have legislative duties to environmental impact, and these duties, normally, depend on the size of your company.

The obvious item that we all have in common is packaging. It is very easy to say that all packaging should be from recycled materials, but do not forget that all of your materials must be bio-burden and animal product free at the point of packaging. If the material is recycled how do you know its source? It is unlikely that any sterile packs will be made from recycled materials, purely because of this fear.

However it is also a truism that it is difficult to source metals that have not been recycled in some form; they have, after all, been recycled from rocks. One thing we can be sure of though is they have not been recycled from medical use as hospitals, etc., dispose of contaminated materials diligently.

So as medical devices designers what can we do? Certainly exterior packaging can be recycled after transportation. Devices that are supplied nonsterile can also have their packaging recycled after transportation. If our risk analysis says it is safe to use recycled materials in certain conditions (and we have avoided contamination from animal by-products, etc.) then there is no real reason why it cannot be done. But remember, if you are doing something new you will need to do a thorough evaluation to show what you are doing is safe (see risk analysis and evaluation chapters later).

Also those of you who deal with instrumentation (say, ultrasound imaging devices) are required to recycle under the European WEEE (Waste Electrical and Electronic Equipment) directive. However, even this is fraught with issues. In the USA the EPA (Environmental Protection Agency) considers some electronic components as hazardous (cathode ray tubes for example), so even this avenue is unclear for the designer.

Thus, I think you can see we are at a bit of an impasse. Those of us designing devices for direct clinical use would love to be "greener" but, for obvious reasons, our hands are tied. Hence we can only do "what is *practicable*." Those of us designing medical devices for the domestic/consumer market have more freedom; but it must be remembered that these are still medical devices and hence you are still bound by medical device regulations.

Many of you will be designing and selling reusable devices (such as forceps). These are amongst the greenest of medical devices as they get used over and over again. They are not, strictly, recycled; they are reused. Again, even this is becoming cloudy as many online articles now state things such as "reusing a medical device is like buying a secondhand toothbrush." We will never escape the hyperbole of the press and, even worse, the Internet. In Europe it is now commonplace to be required to justify why your device is single use (in order to stop companies from stopping items being reused to drive up income streams). This clearly flies in the face of the previous public concerns. Another paradox is that many hospitals now want single use items to be separately packaged and supplied sterile – this obviously saves them sterilization and inventory costs but is a terrible waste of packaging materials.

Hopefully you will begin to see that this subject (for medical devices) is a paradox. On the one hand we have guidelines and legislation to force us to recycle, but on the other hand we have legislations and guidelines that force us not to recycle. The only thing we can do is "what is reasonably practicable." Table 8.12 is designed to help you do this. But do not forget your device is part of a system; so you must look at the system as a whole.

As you should now see, if you develop your PDS well then most of these "seven wastes" should have been covered from day one. But you should reexamine them in the light of ever-increasing environmental legislation – especially those of you with electronics in your devices.

One thing to remember: if you save material, energy, or transportation (i.e., any of the "green" costs), then your main device is cheaper to manufacture and distribute. Hence you can reduce sales cost or your margins will increase. So do not think that all "green" activities must "cost" – they are more likely to "save."

8.8 Design for Usability (DFU)

I have separated this from the other DFX partners, mainly because it is uniquely important but also because I have invented it as a collection of three subdisciplines: ergonomics, man-machine interface, and desirability. All three work together, as they highlight the fact that your device is to be used by someone, and that someone has to be able to use it!

Some fundamentals to consider:

- Who is going to use the device?
- Who is going to install the device?
- How much training is required to assemble/use the device?

- Is the device going to be moved? If so, can it be moved with ease?
- Where will the device be stored when not in use?

Do you remember the "5 Whys" I introduced earlier? Perhaps now is the time to introduce the "5 Whos":

1. Who is going to use it?
2. Who is going to provide training?
3. Who is going to assemble the device?
4. Who is going to transport the device?
5. Who is going to store the device?

Let us look at the first "who" in detail. Who is going to use it? Who is (or are) the end-user(s)? Once you have identified these then you can consider all of the human interaction that will occur.

Let me give you an anecdote that may or may not be true, but is totally imaginable. On a visit to a nuclear power station an inspector noticed a handle from a public house attached to the main control desk. "What is that?" he asked in alarm. "That," said the technician "is the emergency shutdown button." The technician then proceeded to remove the handle to reveal two identical rocker switches next to each other. Apart from a small sign they were indistinguishable – and in the heat of a shutdown a mistake could easily be made leading to catastrophe. The technical team added the beer-pull so that they could never hit the wrong switch. The designer of the panel had got it wrong!

Table 8.13: 6σ Seven Wastes Applied to DFU

Waste	Description/Question
1. Waste of overproduction	• Are there lots of spare items that will not be used?
2. Waste of waiting	• Are the users waiting for things to happen without any information about what is happening?
	• Do the end-users need to order anything specific?
3. Waste of transporting	• Does the device require significant transport for calibration, sterilization, etc.?
4. Waste of inappropriate processing	• Have you thought about the MMI? • Are there numerous subtasks to achieve an overall goal?
5. Waste of unnecessary inventory	• Do your operating/procedure packs come with overly numerous spare items? • Have you discussed inventory with the end-user?
6. Waste of unnecessary motion	• Are the end-users' eyes having to dart all over the place? • Are the end-users' hands, eyes, and arms involved? If so have you considered the ergonomics?
7. Waste of defects	• Is your process complex enough to cause procedural mistakes that do not cause harm but create waste that will annoy your end-user?
"New" Waste	
7(c) Waste of untapped human potential	• Have you sought advice from those who may understand the usual processes in situ?

Figure 8.24
Comparison of two knobs: Which is better (a) or (b)?

The previous anecdote should reveal to you that considering the end-user means "*considering the end-user*"; there are two meanings of consider – to just think about something for a decision, or to think about someone with consideration. These are, in effect, all of the potential failure modes when in use. By now you should have seen that this is really an extension of the FMEA we saw earlier. However, instead of just thinking of how the device can fail operatively, we also consider how it can let the end-user down.

8.8.1 Ergonomics

Ergonomics is the science associated with the ability of things to fit with the human form. There are two considerations: does it "fit," and do the forces required to perform the function fall within normal human ranges. Consider your desk at the office. The chair can be designed so that it feels comfortable to sit in (i.e., passes the "it fits" test), but are you sitting in such a position that when you work at your desk it causes you to strain to reach things (i.e. it fails the "within normal ranges" test)? Your biomechanics reference books will come in handy again.

One of the main considerations is lifting. If your device needs carrying at any time can it be lifted? If so how many people are required to lift it safely? Does it need handles? Virtually every country has health and safety laws associated with lifting and carrying; your device must be designed with due consideration of these.

Consider Figure 8.24. Here we have two knobs to examine. Which is better? It all depends on what they are to be used for. Certainly if the knob was for turning then the profile of (a) and its indicator dot makes this more usable and ergonomically more suitable. Knob (b) on the other hand is far more suitable for a "pulling" action. Hence you tend to see knobs of type (a) on amplifiers, etc. where they need to be turned; you tend to see (b) on drawers and cupboards (and old church organs) where they need to be pulled. What if the end-user was infirm – could they activate the device now? Do they need anything special?

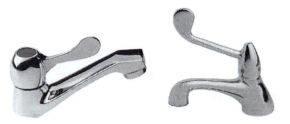

Figure 8.25
Common "lever" type water taps (faucets).

The lesson here is do not reinvent the wheel – many of the ergonomic lessons are in textbooks catalogs, and your end-users' brains.

Figure 8.25 illustrates two typical, modern water taps (or faucets). For people who cannot grip easily, the standard tap is better than useless. These "lever" types of tap have been in use in hospitals for years (because you don't need to use your hands to activate them); they are therefore perfect for those who cannot use their hands. Modern electronics means that the taps can now be automatic. This is an example of "inclusive design": designs that can be used by able and physically challenged people alike. Inclusive design is a natural extension of ergonomics; it expands the population to include those who were previously excluded.

There is a plethora of books on ergonomics and inclusive design. I leave you to find one that you like. Just remember you cannot design the item to be ergonomic if you have not found out "who" it is to be ergonomic for! Please do not forget that your device will eventually be applied to a patient: hence they MUST be in the list of "whos."

8.8.2 Man-Machine Interface

This crosses over with ergonomics. Its main concern is the interaction between your device and any of the end-users. It is used extensively where computers are involved and many times, mistakenly I propose, is only applied to web pages (and the like). However, you should consider it with greater detail.

Consider a piece of equipment to be used in the operating theater, where there is lots of blood. Blood and other body fluids are great lubricants and make items very slippy. If your device has been coated with body fluids, are any knobs, switches, etc. "turnable"? Can the theater staff (wearing surgical gloves too) actually use your device?

If your device relies on a software interface, is it intelligible by the user, not just the program author? You probably have experienced a new mobile phone and the complexity for the first few uses.

Hence MMI is all about making any interface with an end-user logical and easy. Once again, using the FMEA sheet from earlier will help you to solve this issue too. After all, not being able

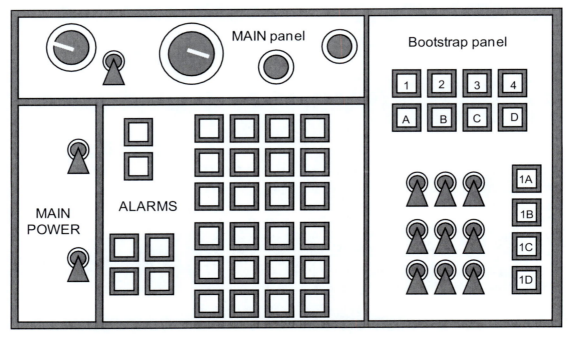

Figure 8.26
A daunting man-machine interface.

to flick a switch because it is covered in blood is just as much a failure as the whole machine going up in smoke (remember I proposed to you the concept of damaging your brand!).

Consider the layout illustrated in Figure 8.26. Those of you designing devices such as ultrasound imaging systems will be used to arrays of buttons like that in Figure 8.26. The buttons, knobs, dials, and switches may be ergonomic, all fit for purpose and all easily activated (even when covered in body fluids). However, does the layout make you stand back and a small voice in the back of your head say "Oh my God"? MMI is all about removing the "OMG" reaction. Yes, some units are complex by their very nature, but one does have to think about the end-user!

8.8.3 Design for Desirability

We now go back to the discussion about design in Chapter 1. Desirability breaks down to "I know I need it, but do *I* want it?" Here I will use one example to demonstrate what I mean. If you are short or long sighted then you will probably have some spectacles somewhere. Some of you will have contact lenses because you do not want to wear spectacles. Those of you who wear spectacles probably took some time choosing the right frames; and why not, they will be attached to your face for quite some time. If we take so much effort over spectacles why do we not do the same for other medical devices? Why do we neglect the poor patient who has to have this "thing" with them all day, maybe forever?

If there is one thing I hope to get across in this book, it is the concept of "healthcare jewelry." All of our medical devices should not just function – they should look good too. Something I learned quickly was that if you want to command a high price your device had better look expensive! Because consumer items (such as DVD players, etc.) are so cheap, technology alone no longer commands high added value. Hence revisit Chapter 1 and reread the quote from a certain Steve Jobs. No one could argue that Apple computers ignored this concept – neither should you.

Some of you may not have the skill to make your device "desirable." However there are numerous product designers and industrial designers who can help. They will charge you, but as I was told some time ago "only one company can be the cheapest, competing on price alone is not enough – be different by design." If your raison d'être is to be the cheapest, then so be it. However if you want to be the market leader I suggest you take the meaning of this sentence to heart, just as Apple did…and it did them no harm.

8.9 Summary

In this chapter we examined the tools required to perform a satisfactory detailed analysis. Hopefully you will have seen that having a well written PDS takes all of the hard work out of design!

We examined specific team selection and the importance of the team/design lead in the team's overall performance. We also examined their role in relation to quality management. I then introduced standard design documentation to ensure that QM is robust.

We then looked at specific tools that will make your life easier, such as computer-aided design and computer-aided analysis. We examined the role of various visualization and rapid prototyping techniques and saw how they can help you to produce a design worthy of the problem. I then introduced you to the family of DFX tools and showed you how they can be used to minimize your exposure when risk analysis, necessary for all medical devices, is conducted.

References

Ashby, M. F. (2004). *Materials selection in mechanical design*. Oxford: Butterworth-Heinmann.

Boothroyd, G., Dewhurst, P., & Knight, W. A. *Product design for manufacture and assembly*. CRC Press.

Grimes, S., Donaldson, J., & Gomez, G. C. (2008). *Report on the environmental benefits of recycling*. Bureau of International Recycling.

Tuckman, B. W. (1965). Developmental sequence in small groups. *Psychological Bulletin*, *63*(6), 384–399.

Tuckman, B. W., & Jensen, M. A. C. (1977). *Stages of small-group development revisted*. Group and Organisation Studies. pp 419–427.

Williams, D. F. (1999). *The Williams dictionary of biomaterials*. Liverpool University Press.

Williams, D. F. (2003). Revisiting the definition of biocompatibility. *Medical Device Technology*, *14*(8).

## CHAPTER 9

Evaluation (Validation and Verification)

9.1 Introduction

We now arrive at the pinnacle of our design activity. Everything we have done, so far, has led to this day…the day we see if our design works. The recent changes to medical device legislation in Europe and tightening of existing guidelines elsewhere has made evaluation (of medical devices before they are authorized for sale/use) ever more important. In the past medical device designers were almost derisory in their attention to this aspect of the design process. They left this part to someone else as if it were not their problem – well *it is!* It is now impossible for you to ignore this important aspect of the design cycle; it is so important that it has its own section in the PDS called Acceptance Criteria.

In the EC, this aspect is known as *clinical evaluation*; under FDA guidelines it is called *validation and verification*. Verification means to compare design input to output; validation means ensuring the device meets clinical requirements, is safe to use, and does what it is supposed to (in a clinical sense). As I said in Chapter 4, I will call the whole *evaluation*.

There are two common forms of evaluation: *in vitro* and *in vivo*. The former means in a laboratory, the latter means on living subjects. So, for example, a test that shakes a device to pieces would be in vitro. A literature review selecting and confirming precedence would also be in vitro. A test that counts how many times a patient uses a device per day would be in vivo. We shall see how these two exhibit themselves in real life.

9.1.1 Clinical Trial or Clinical Evaluation: What Is the Difference?

There is a very important distinction to be made between these two. Firstly, which is higher up the food chain? It is simple: a clinical trial is always a part of an overall clinical evaluation. You cannot release a product without conducting a clinical evaluation, but you may have no need to conduct a clinical trial. It is important to note that most countries now state that for some devices clinical trials are a mandatory. For example, in the EC any Class III device must have a clinical trial as a part of its evaluation. You must check for yourself what the current regulations are in the states in which you wish to sell.

Clinical evaluation takes place prior to any release; this inherently means that anything undergoing a clinical trial is a nonapproved product. In EC terminology it has no CE mark; in the USA it has no 510(k). Hence a clinical trial, by definition, is conducted using a

Medical Device Design.
DOI: http://dx.doi.org/10.1016/B978-0-12-391942-7.00009-X

product that is not for sale – one which has no approval to be used on the general public. A clinical trial is always conducted in vivo – on human subjects and under highly controlled circumstances. Therefore clinical trials are highly regulated; special approval must be sought and they are very expensive (you would be lucky to have any change from $150,000 for a small trial). Hence it is important that you:

  i. Avoid the term *clinical trial* unless you are actually intending to conduct one.
 ii. Only start a clinical trial if you are forced to do so by regulation or by lack of evidence to suggest it is unnecessary.
iii. Obtain formal approval from your regulatory body to go ahead.

We shall examine clinical studies and trials in more detail later in this chapter.

9.1.2 Why Do an Evaluation?

There is one major overriding consideration: patient and end-user safety. You must ensure your device does what is supposed to do, does it in a way that meets any regulatory requirements, and does it safely. For all medical devices you must be able to make a statement called a *declaration of conformity*; how can you make such a statement if you have not done a full and thorough evaluation?

Hence the need for an evaluation is quite simple; it is a regulatory requirement and no design is complete without one.

9.1.3 What Is in an Evaluation?

Hopefully, you will have remembered that to meet ISO 13485 an *evaluation* (or *validation/ verification*) *procedure* is required. This procedure must cover the steps as stipulated by the pertinent regulatory body, but as they are all asking for pretty much the same thing the procedure can be general. At the end you must produce a report detailing what you have examined and making the statement that your device has passed the evaluation – hence making a declaration that it conforms to whatever regulation you are working to.

In the USA the medical device 510(k) application process virtually forces you to go down this path. In the EC the new MEDDEV guidance documents also force you to follow a particular path. Hence, as per usual, you should have the 510(k) application guidance (510k sections 12–20) and the MEDDEV 2.7.1 rev 3 documents at hand; and as usual the relevant website is the point of source.

Clinical evaluations can seem daunting and confusing. I am going to try and simplify them by showing you that if you use "design" as your guide then everything will fall into place.

9.1.4 Relationship with the PDS

As with all previous items your first port of call should be your PDS. Hopefully you will remember that there is a specific PDS section for evaluation: your acceptance criteria. This section states what your device has to be measured against before it can be accepted for use. If your PDS has been written properly then planning your clinical evaluation is easy.

9.1.5 Method of Demonstration

In all cases this entails the production of an evaluation report. I will try and make this description viable for both the USA and EC regulations. In essence the only real difference between the two systems is the wording; all will become clear. The report will need to contain specific sections, as described subsequently. The following sections of this chapter do not necessarily map to any particular section of a 510(k) application or an EC technical file as they are all applicable to all sections. You need to decide which evaluation tool is best to measure your particular outcome.

9.2 Risk Analysis

This is the "Daddy" of them all. It is the precursor to all that follows. In essence a clinical evaluation follows an overall risk analysis – however in practical terms they are so interlinked that it is impossible to separate them. Hence I am introducing *risk analysis* as a tool to be used in your overall evaluation of your device. In fact, an overall risk analysis is essential for all medical devices, in the USA or in the EC. So this is the right time to think about it.

Unlike many other disciplines we do not have any choice about risk analysis; we must use the ISO standard "BS EN ISO 14971:2009: Medical devices. Application of risk management to medical devices" (BSI, 2009). The first (and largest) portion of the document concerns risk management. You must have a risk management procedure, and this standard actually gives it to you – so there can be no mistake. A subsection concerns risk assessment; again this process is almost handed to you. The only thing you can change is how you present your findings; all else is legislated. Avoid using this standard at your peril. Indeed if you avoid it in the EC you will lose your CE mark and your right to be a medical device manufacturer; do not think the FDA will be any less stringent!

The important thing to consider with risk analysis is the simple scientific principle of cause and effect. In risk analysis we do this backwards; we think of the effect(s) and then determine the root causes(s). However, unlike the design FMEA presented earlier we are only asked to consider effects that can harm a patient, an end-user, or a bystander. Any embarrassment to the company is a secondary issue, if that. Thus we imagine horrible things that could happen as a result of your device being used and then determine the associated risk. But we do need to start thinking of "nasty things" that can happen.

The essence of ISO 14971 is that you must have a risk management procedure in place. It is a nice standard. It not only gives you the procedure (so there is no need to develop your own, all you need do is make it work for your company), it gives you sample risk analysis forms too. So I now intend to present these to you; but they should not be a surprise as they are very similar to those we saw in FMEA.

9.2.1 Identifying Risks/Hazards

This aspect of the clinical evaluation process is to ascertain risks and hazards. The first port of call is Annex C of ISO 14971 (see Appendix C for a full table). Table 9.1 illustrates one small subsection of Annex C. The whole table, if worked through, helps you to identify risks pertaining to your device.

The whole aim of Annex C is to get you thinking about risks when the device is in the real world, as opposed to in your design office. You should consider all aspects related to your device. You should put yourself in the place of use. If you are unable to imagine this then you need to gain some experience related to the place of use or get hold of someone who has.

Some of the questions posed by Annex C will not be related to your device. If so, simply state N/A and then state why in the Comments column. Also, you will find that some sections repeat themselves. This is perfectly reasonable – the repetition may have occurred so make sure you look at all angles. When completing this table consider issues at the hospital, at your suppliers, and in your warehouses. You need to use these questions to think of *any* potential risk – no matter how negligible it may seem. It is only when we come to do the analysis that we consider the level of risk.

Table 9.1: Subsection of Annex C ISO 14971

Subsection	Applicable/Not Applicable	Comments
c.2.8 Is the medical device supplied sterile or intended to be sterilized by the user, or are other microbiological controls applicable? Factors that should be considered include: 2.8.1. Whether the medical device is intended for single use or reuse packaging 2.8.2. Shelf life issues 2.8.3. Limitation on the number of reuse cycles 2.8.4. Method of product sterilization 2.8.5. The impact of other sterilization methods not intended by the manufacturer		

CASE STUDY 9.1

Consider a single use device that is supplied nonsterile and relies on the end-user using a steam sterilization process before use. Using Table 9.1 consider any potential hazards.

Subsection	Applicable/ Not Applicable	Comments
c.2.8 Is the medical device supplied sterile or intended to be sterilized by the user, or are other microbiological controls applicable?	✓	Single use device to be steam sterilized before use.
Factors that should be considered include:		
2.8.1. Whether the medical device is intended for single use or reuse packaging	✓	Can it be mistakenly used nonsterile? Can it be mistakenly reused? Does the packaging make its sterility status obvious?
2.8.2. Shelf life issues	✓	Will the device deteriorate over time on the stock shelf, in transit, or in the hospital? Once sterile how long will the device remain sterile and in what conditions?
2.8.3. Limitation on the number of reuse cycles	✓	Will resterilization (due to not being used) cause issues? How many times can it be resterilized without having detrimental effects?
2.8.4. Method of product sterilization	✓	Can the device be washed/cleaned in normal clinical machines? Do we have a certificate stating it is sterilizable? Have we checked that it can be sterilized using normal clinical methods? Have we checked that the sterility conforms to standard procedures in all states in which it is being sold? Will the device store heat and hence be able to burn/scald patient end-users?
2.8.5. The impact of other sterilization methods not intended by the manufacturer	✓	What happens if the device is ETO sterilized? What happens if it is irradiated? Do either have detrimental effects? Do either cause any issues related to usability?

Once you have identified related areas (using Annex C), you will also need to think of the appropriate hazards to be inserted in the Comments column. In most cases this is like asking "How long is a piece of string?" However, ISO 14971 comes to the rescue again. Annex E helps us to imagine particular issues. Although Table E2 is useful, for the first-time user it

Table 9.2: Examples of Hazards

Examples of Energy Hazards	Examples of Biological and Chemical Hazards	Examples of Operational Hazards	Examples of Information Hazards
Electromagnetic energy *Electricity* Line voltage Is it connected to mains supply? Leakage current - enclosure leakage current - earth leakage current - patient leakage current Is it DC or AC? Is it single phase or three phase? *Electric fields* Will it produce a magnetic field? Can it be affected by magnetic fields? Data contamination through interference? EMC compatibility? *Light* Does it emit light? Can it cause damage to eyes? Will it cause temporary blindness (flash)? Does it need to be used in a dark or light environment? Is it a laser? **Radiation energy** *Ionizing radiation* Is there any? Is it directional? How much? *Nonionizing radiation* UV sunburn?	**Biological** Bio-burden? Bacteria? Viruses? Other agents (e.g., prions) Re- or cross-infection? Animal-based products? Any of the above due to reuse? **Chemical** Any acids or alkalis? Any processing residues? Any contaminates? Any additives or processing aids? Any cleaning, disinfecting, or testing agents? Can any of the above cause degradation? Will it use/transmit any life-threatening chemicals (e.g., medical gases, anesthetic products)? What effects can any of the above have on the device itself? **Biocompatibility** Toxicity of chemical constituents, e.g., - allergenicity/ irritancy - pyrogenicity	**Function** Effects due to: Incorrect or inappropriate use? Incorrect measurement? Erroneous data transfer? Loss or deterioration of function? Misuse? Ignoring a warning or error message? Not checking functionality before starting? **User error** Effects of: Lack of attention? Forgetfulness? Lack of training? Ignoring the rules? Lack of knowledge? Is there any assembly that could go wrong?	**Labeling** Are the instructions for use adequate? Are the indications clear? Contraindications clear? Are the performance criteria clear? Are the above written for all end-users? **Operating instructions** Written with the end-user in mind? Inadequate specification of pre-use checks? Overcomplicated operating instructions? The effects on the device if any of the above occur? **Warnings** Any side effects? Any hazards likely with reuse? **Specification of service and maintenance** Any special service instructions before reuse? Anything needs disposal before reuse? Any pre-use checks?

(Continued)

Table 9.2: (Continued)

Examples of Energy Hazards	Examples of Biological and Chemical Hazards	Examples of Operational Hazards	Examples of Information Hazards
Thermal energy Ductile-brittle transition? Burning/scalding? Radiated, conductive, or convective heat? Freezing? Will it act as a heat sink? Will it excessively heat or cool the environment? **Mechanical energy** *Gravity* – Can it fall? – Can it topple? – Can suspension fail? *Vibration* Can it affect the user? Can bits become loose? Will it produce excessive noise? *Stored energy* Can it spring back? Are there clips that can pinch? *Moving parts* Can clothing get caught? Can fingers get caught? *Torsion, shear, and tensile force* Have you considered excessive loading? *Moving and positioning of patient* Will the patient need to be moved?			

(*Continued*)

Table 9.2: (Continued)

Examples of Energy Hazards	Examples of Biological and Chemical Hazards	Examples of Operational Hazards	Examples of Information Hazards
Acoustic energy – ultrasonic energy – infrasound energy – sound *High pressure fluid injection* Will it "inject"? Can it cut? Will it overinflate?			

(Modified from ISO 14971:2009)

is meaningless. Hence I have taken this table and converted it into questions you should ask yourself. This table is by no means complete; it is only a starter and you can use this to build your own, more detailed, list of hazards.

As with earlier aspects of quality in design, it is worth using the following W questions:

> Whom: Hazardous to whom? The patient? The end-user? Other devices!
> What: What makes this a hazard?
> Why: Why is it hazardous? (If this is not obvious, you may have to describe in more detail. Things like electric shock or scalding need no expansion – apart from the potential degree.)

I have not completed the table, perhaps you would care to complete it?

You should foresee that this will be a lengthy process. It is very time-consuming and results in loads of paperwork. But by the end you will have thought of just about any stupid thing that could be done by every possible end-user. When you consider that you are about to release a medical device onto an unsuspecting world – that may just kill someone – then this exercise is highly worth it. In my experience it only takes a few days, but it is a few days well spent. Remember, risk analysis is a mandatory exercise, so you may as well do it right! The other thing to remember is that if you have written your PDS in the first instance, then all of these risks will have already been mitigated!

9.2.2 Assessing Level of Risk

The similarity between FMEA and the ISO 14971 risk analysis form is no coincidence. However they depart in one major aspect: FMEA was used to design out risk whereas the risk analysis is to check whether any residual risks remain. If you designed your device using a well constructed PDS then this analysis should return a "safe to use" result. What does "safe to use" mean? It simply means that any risk of use is outweighed by the clinical benefits.

CASE STUDY 9.2

A single use device is to be steam sterilized at the point of use using a desktop sterilization unit. This unit also performs washing. Using Table 9.2 determine any potential hazards that may be part of your risk analysis.

Potential Hazard	Whom?	What?	Why?
Electricity			
Line voltage	Staff	Electric shock	Death Burn
Leakage current			
– enclosure leakage current	Staff	Electric shock	Death
– earth leakage current	Staff Patient	Electric shock Abrupt end of treatment	Potential for injury
	Other devices	Abrupt shutdown	Potential for injury
Is it DC or AC?	Device	Breakdown	No sterilization thereafter
Is it single phase or three phase?	Device Staff	Breakdown Electric shock	
Electric fields			
Will it produce a magnetic field?	Other devices Patient	E-Mag field effects E-Mag field effects	Interference Attracting metals Affecting pacemakers
Can it be affected by magnetic fields?	The devices	E-Mag field effects	Interference Attracting metals
Data contamination through interference?	Other Devices	E-Mag field effects	Interference
EMC compatibility?	Other Devices	E-Mag field effects	Need for EMC test
Thermal energy			
Ductile-brittle transition? Burning/scalding? Radiated, conductive, or convective heat? Freezing? Will it act as a heat sink? Will it excessively heat or cool the environment?			

Potential Hazard	Whom?	What?	Why?
Biological			
Bio-burden? Bacteria? Viruses? Other agents (e.g., prions) Re- or cross-infection? Animal-based products? Any of the above due to reuse?			
Labeling			
Are the instructions for use adequate? Are the indications clear? Contraindications clear? Are the performance criteria clear? Are the above written for all end-users?			

For example, we all know that x-rays are an ionizing radiation and hence pretty dangerous things. After all, if Marie Curie had the benefit of hindsight she may not have carried isotopes around in her pocket! However, how would modern medicine get along without the x-ray machine? How would your dentist examine your roots without this device? While it is impossible to remove all of the risk, we are able to reduce it to levels where the benefits outweigh the risk. As such all hospitals, all dentists, and all veterinary practices have an x-ray machine.

The same argument must be applied to your medical device. You must be able to prove, using risk analysis, that "*the medical benefits outweigh the risk.*"

In order to be able to make this statement, we must consider the risks/hazards presented in the previous section; and then for each one identify the root cause or (if applicable) the root causes. To illustrate this Figure 9.1 is a typical FMCA pro forma; this has been modified from ISO 14971 to coincide with terms we have already met.

Use of Annex C will identify potential failures/hazards. The relevant section number of Annex C is entered into box 1 (see Figure 9.1 for numbering of boxes). You may have a number of different

RISK ANALYSIS						
Characteristic	1	Life Cycle Phase	Design, Manufacture, and Supply	Comment		
Failure/Hazard	2					
Effect	3					

Root Cause(s)	Hazard Relevance ✓ = relevant				Estimation of risk: Likelihood x Severity at start of life cycle phase L x S = RPN			Risk reduction activities (gray/black zones) (if practical): Comment if required (white only)	Estimation of **residual risk** at the end of this life cycle phase			Can the risk be reduced further? (gray)	Action proposed to protect from residual risk, implementation and verification of efficacy?	Additional hazards introduced by risk control measures? If so, what action taken?
	Patient	User	Bystander	Environment	L	S	RPN		L	S	RPN			
4	5				6			7	8			9	10	11

Figure 9.1

Example risk analysis pro forma.

failure modes for a particular hazard; a single failure mode is entered into box 2. Each failure mode/hazard will have a particular effect (note this is related to the patient, the end-user, or the surrounding environment – box 5); this is explained in box 3. Now you have to identify the root cause, or the root causes (box 4); it is very likely that each hazard/failure will have a number of potential root causes. This is where our Ishikawa diagrams and our reliability calculations come into force (Chapter 7).

Table 9.3 is a summary of the potential root causes as stated in ISO 14971. It is by no means comprehensive but it gives you some ideas. As stated earlier, and continuously throughout this text, a comprehensive PDS and design procedure will have anticipated all of these root causes and designed them out!

For each cause we now assess the risk (box 6). Similar to what we examined in FMEA, we determine a level of severity (S) and a likelihood of occurrence (L). But unlike FMEA we DO NOT include delectability. Our assessment of RISK level (RPN) is

$$RPN = S \times L \tag{9.1}$$

As with FMEA, we need guidelines on setting values of L and S. ISO 14971 suggests values but also lets the company allocate their own appropriate levels; those in Table 9.4 are commonplace.

Note that severity is based on potential for injury; company embarrassment is no longer a consideration!

We now have to assess if the risk is acceptable or not. Table 9.5 illustrates a typical risk evaluation table. ISO 14971 allows you to devise your own threshold values, but it is very common to have three zones: a low risk zone (no controls required); a medium risk zone (controls should be examined); and a high risk zone (controls need to be implemented to reduce risk).

Table 9.3: Example of Root Causes

General Category	Examples of Causes
Incomplete requirements	Inadequate specification of: – design parameters – operating parameters – performance requirements – in-service requirements (e.g., maintenance, reprocessing) – end of life
Manufacturing processes	Insufficient control of changes to manufacturing processes Insufficient control of materials/materials compatibility information Insufficient control of manufacturing processes Insufficient control of subcontractors
Transport and storage	Inadequate packaging Contamination or deterioration Inappropriate environmental conditions
Environmental factors	Physical (e.g., heat, pressure, time) Chemical (e.g., corrosions, degradation, contamination) Electromagnetic fields (e.g., susceptibility to electromagnetic disturbance) Inadequate supply of power Inadequate supply of coolant
Cleaning, disinfection, and sterilization	Lack of, or inadequate specification for, validated procedures for cleaning, disinfection, and sterilization Inadequate conduct of cleaning, disinfection, and sterilization
Disposal and scrapping	No or inadequate information provided Use error
Formulation	Biodegradation Biocompatibility No information or inadequate specification provided Inadequate warning of hazards associated with incorrect formulations Use error
Human factors	Potential for use errors triggered by design flaws, such as – confusing or missing instructions for use – complex or confusing control system – ambiguous or unclear device state – ambiguous or unclear presentation of settings, measurements, or other information – misrepresentation of results – insufficient visibility, audibility, or tactility – poor mapping of controls to actions, or of displayed information to actual state – controversial modes or mapping as compared with existing equipment – use by unskilled/untrained personnel – insufficient warning of side effects – inadequate warning of hazards associated with reuse of single use medical devices

(Continued)

Table 9.3: (Continued)

General Category	Examples of Causes
Failure modes	– incorrect measurement and other metrological aspects – incompatibility with consumables/accessories/other medical devices – slips, laps, and mistakes
	Unexpected loss of electrical/mechanical integrity Deterioration in function (e.g., gradual occlusion of fluid/gas path, or change in resistance to flow, electrical conductivity) as a result of aging, wear, and repeated use Fatigue failure

(Source: ISO 14971:2009)

Table 9.4: Example Table of Severity Levels

	L		S
5	frequent 1/100 uses or	5	catastrophic Death
4	probable 1/1000 uses or once per week	4	critical Major injury (loss of limb, etc.): life-threatening injury
3	occasionally 1/10,000 uses or once per quarter	3	serious Minor injury requiring treatment
2	remote 1/1,100,000 uses or once per year	2	minor Minor injury NOT requiring treatment
1	improbable 1/1,000,0000 uses or once every 3–5 years	1	negligible Minor irritant to patient or end-user

Box 7 is reserved for any description of remedial action (or comments if in the white zone). A new level of RPN should be determined and entered into box 8. If this is insignificant all is fine; if, however, the risk is still significant then the last next two boxes need completing. You need to examine if the risk can be reduced any further. If it can then this needs to be described in box 10. Simply speaking, if the residual risk is unacceptable then you must go back to the drawing board – but if you have a good PDS this eventuality should not happen. If the risk is significant you must assess if the residual risk is outweighed by clinical benefit and you may need to instigate further controls.

Box 11 is simple: changes you have made to reduce risk may have a knock-on effect. You may have inadvertently introduced new risks. This box forces you to look at that outcome.

Once you have completed a form for each hazard you will need to produce a risk analysis report, one that contains each and every completed FMCA table – and you will have many. These individual forms are collated and together they define whether your device has any residual risk that is not acceptable. The front page of this report summarizes this statement, but someone competent must sign it of and the "sign off" must contain a statement confirming that the clinical benefits of the device outweigh any risks due to its use.

9.3 Criteria-Based Evaluation

It is very important to recognize that a *full* evaluation of your design is mandatory in both EC legislated countries and under the FDA. Furthermore it should be recognized that many issues highlighted in the risk analysis, described previously, cannot be addressed without performing some form of evaluation on the device itself. The recent version of the European Medical Devices Directive has made a clinical evaluation mandatory, see MedDev 12.2/6 for more information (EC, 2010). The FDA has guidance in their "Control of Design" guidelines for manufacturers (FDA, 1997). In all cases you have an obligation to show that your device meets both your design inputs and the requirements to be called a medical device. Just performing a controlled design process without this final stage is not enough. Furthermore,

---

CASE STUDY 9.3

During the manufacture of a hypodermic syringe it was identified that some material may have originated from a warehouse in Japan. Assess the risk of this potential hazard.

Any FDA registered organization would have received an official letter from the FDA in 2011. This letter requested that the organization check that no materials had been sourced from Japan; this letter was specifically concerned with the nuclear reactor failure that followed the 2011 tsunami and hence potential radioactive contamination of any materials.

What is the potential hazard?

From Annex C, though this is debatable, the potential issue is 2.4.3:

> *2.4 What materials or components are utilized in the medical device or are used with, or are in contact with, the medical device? Factors that should be considered include:*
>
> *2.4.3. Whether characteristics relevant to safety are known*

From Table 9.2 the hazard is clearly ionizing radiation. The effect is injury to the patient, to the end-user, and possibly to the environment. Hence there are two potential effects:

1. Injury to the patient and/or end-user.
2. Contamination of the storage environment, which in turn can lead to injury to the end-user(s).

RISK ANALYSIS												
Characteristic	2.4.3			Life Cycle Phase	Design, Manufacture, and Supply		Comment		FDA request letter ref x.y.z			
Failure / Hazard	Supply of radioactive components due to potential contamination in Japan											
Effect	Ionization injury to patient											
Root Cause(s)	Hazard Relevance ✓ = relevant				Estimation of Risk: Likelihood × Severity at start of life cycle phase L × S = RPN			Risk reduction activities (gray/ black zones) (if practical): Comment if required (white only)	Estimation of residual risk at the end of this life cycle phase			Can the risk be reduced further? (gray)	Action proposed to protect from residual risk; implementation and verification of efficacy?	Additional hazards introduced by risk control measures? If so, what action taken?
	Patient	User	Bystander	Environment	L	S	RPN		L	S	RPN			
Suppliers inadvertently use material(s) that have passed through Japan post April 2011	✓	✓	✓	5	4	20		Contacted all subcontractors to check providence of all materials	1	4	4	No	Written confirmation obtained from all suppliers that no materials originate from or passed through Japan	No additional hazard

This case study has demonstrated how, even after a device has been placed on the market, a thorough risk analysis can be used to check if anything needs to be done. In this case the analysis showed that we could not be sure that the materials were not contaminated, hence a likelihood of 5. However, after contacting all suppliers for their materials' providence it was apparent that only a random mishap would mean the use of contaminated materials, hence L = 1. Notice the action to ensure that the risk is controlled; requesting letters of providence from the suppliers will ensure they keep their eyes on the ball too!

the outcome of your evaluation has to be formally signed off by a qualified person and then placed in your device's technical file (or DHF). There is little doubt that this chapter should be governed by regulatory documents. However they tell the reader what to achieve, but not how to get there. They state that you should demonstrate the benefits to the patient's treatment, the patient, and the end-users…but they do not tell you how. This chapter is aimed at addressing that shortcoming.

Table 9.5: A Typical Risk Evaluation Table

S	Negligible: 1	Minor: 2	Serious: 3	Critical: 4	Catastrophic: 5
L					
Frequent: 5	5	10	15	20	25
Probable: 4	4	8	12	16	20
Occasional: 3	3	6	9	12	15
Remote: 2	2	4	6	8	10
Improbable:1	1	2	3	4	5

Key:
Dark: Unacceptable (>10) – Risk must be controlled and RPN reduced.
Gray: Significant (>4) – Risk controls to be investigated to reduce RPN.
White: Insignificant (<5) – RPN need not be investigated further.

We, once again, refer back to our original PDS. Within this document you will have written down your acceptance criteria: what the device must do to be acceptable. Quite often the first stage is to meet the requirements of a specific standard. Whatever the reason, you will have formulated some criteria by which to *measure* success. This section is, therefore, concerned with the introduction of testing methods aimed at proving success.

Almost certain is the fact that you will need to prove that your device does what it is supposed to; this is called *verification*. In medical device terms…does your output meet the demands set by your input? By and large this is conducted in a laboratory and is hence *in vitro*. Sometimes, but rarely nowadays, some animal experiments are required. But they are all concerned with verifying that the device performs as expected.

9.3.1 In Vitro/In Vivo

In vitro literally means "in glass" – this comes from the historical concept of all experiments being conducted using glass test tubes, beakers, and jars. It now covers any experiment that is not, literally, how it is to be used in real life (in our case on a living human being). Experiments on human subjects are called an *in vivo* experiment.

There are two reasons to conduct an in vitro evaluation. The first is to confirm that the performance characteristics of your device are as designed. The second is to experimentally evaluate a failure mode in order to reduce the RPN in a risk analysis.

9.3.2 Accelerated Life Tests

One of the most common in vitro evaluations is an accelerated life test. The actual conditions for the test will come from your PDS and may well be defined by national or international standards. Those of you with sterile packaged items have a mandatory obligation to conduct these, but it is just as important for everyone else. Some common environmental parameters that you may wish to consider are listed in the sections below.

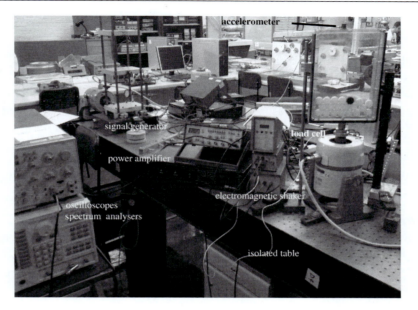

Figure 9.2
Typical vibration test setup. *(Courtesy Staffordshire University)*

9.3.2.1 Vibration

How do vibrations affect your device? Do things become loose (e.g., screws and nuts)? Do components crack or break (fatigue)? Do innocuous components wear through the sterile package? What vibration sources exist in use and in transit?

Once the parameters are identified vibration tests can be easily performed, but specialist equipment is required. Most universities with a mechanical engineering department would have this equipment and some stand-alone companies offer this as a service. Figure 9.2 illustrates a typical vibration test. The normal equipment required is a signal source (normally a single sine wave at fixed frequency and amplitude or a white noise source), a power amplifier, and an electromagnetic shaker. Measurements are conducted using a combination of load cells and accelerometers. The component is mounted, or supported, as it will be in real life and then vibrated for a simulated lifetime. Failures are then observed or measured.

9.3.2.2 Cyclic Loading

Is your device subjected to cyclic loading? Are these cyclic loads mechanical, electromagnetic, or thermal? How many cycles do you expect over the life of the component? By and large this is going to ascertain fatigue life (as it is important we shall look at fatigue in greater detail later).

Once again, specialist equipment is normally required and, as noted earlier most universities would have this equipment in their mechanical engineering or materials technology laboratories. Commonly, the component in question is mounted as close to the real life situation as possible. The cyclic loading profile is devised. This need not be a sine wave; it can be an exact replication of the loading cycle. Running the test system for a stipulated number of cycles tests the life cycle. Time can be accelerated by selecting higher frequencies; but not so high that the test becomes unrealistic due to, say, heat generation or material nonlinearities. Figure 9.3 illustrates two typical cyclic testing machines. Figure 9.3(a) is a low cycle machine (up to about 2 cycles per second); Figure 9.3(b) is a hydraulic system and is high frequency (up to 100,000 cycles per second).

9.3.2.3 Static Loading
The testing machines described in the previous section are normally able to provide static compressive and tensile loads too, as illustrated by the device in Figure 9.4.

9.3.2.4 Humidity and Temperature
Will your device be subject to various levels of humidity? Will it purely be used in a wet environment? Is there likelihood of water absorption? If so, how long does it take to be detrimental? Is there any chance of corrosion? Can corrosion and cyclic loading combine to create a corrosion-fatigue environment?

What is the normal limit of temperature? If your device were to run its full design life at this temperature would it last?

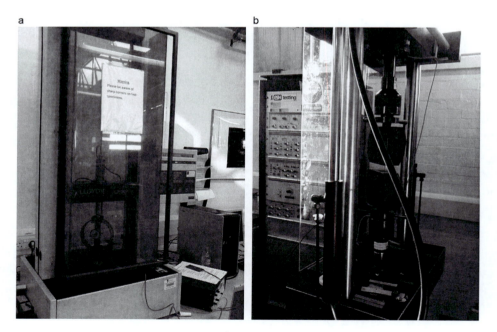

Figure 9.3
Typical cyclic loading machines: (a) electromechanical; (b) hydraulic. *(Courtesy Staffordshire University)*

Environmental chambers are available in a variety of sizes from desktop to those taking up a whole car (Figure 9.5). They can also vary in range and can take temperatures from −40 to +40°C; they can also go from arid to 100% humidity.

9.3.2.5 Normal Use

Apart from extreme use cases, does the device last its design life under normal, average operating conditions? Can the packaging withstand normal transportation? All of the test regimes describe above can be used to examine this aspect of the life cycle.

Figure 9.4
Desktop tensile/compression testing machine. *(Courtesy Staffordshire University)*

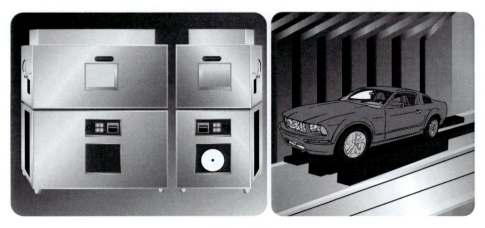

Figure 9.5
Typical environmental chambers: (a) small scale; (b) large scale.

9.3.2.6 Cleaning and Sterilization

Even though your device may be supplied nonsterile you will still have to demonstrate that it can be cleaned and sterilized (if required) by clinical staff. In most cases you will need to supply a certificate to say so. Each country has its own standard cycles for cleaning and sterilization; there is equivalence but you need to be sure that your stated cleaning and sterilization process actually produces a clean and sterile device. While it is tempting to put this into the hands of a commercial sterilization service it is better to have this proved in the actual environment. Hence if your device is to be cleaned and sterilized in the home, that is where you conduct the test…not in a sterile, clean room!

Some things you may wish to consider are:

- Does your device fit into standard washing equipment?
- Does the cleaning process affect your device?
- Do you have any closed holes where detritus can lodge?

9.3.3 Calibration

In many situations, and even before any testing can start, system calibration may be required. In instances where your device makes a measurement, calibration is, most certainly, mandatory. However, the lessons of calibration make the verification of output to input very credible in comparison to any ad hoc methodology. Calibration is a very easy concept if treated with respect; if it is treated in a condescending manner it will come back to bite you.

Calibration is concerned with the referencing of measurements back to international standards of measurement. So, for example, if your device were to measure body weight in kg it should refer to the international standard kg held in France. Clearly you cannot fly to France for every calibration, so each country houses its own standard kg that is referred back to the original. These then produce their own standard kg that is housed in specialist calibration centers based around a country so that they can be used to calibrate against. Hence when, or if, you go to a supplier and purchase a calibrated mass from your supplier, that mass will have a paper trail all the way back to the standard in France. This paper trail is called the *calibration ladder*. The same applies to length, time, etc.

Hence the first thing to learn is

> *you cannot calibrate any item without first having calibrated instrumentation.*

Most ISO 9001 and 13485 companies will have calibrated measurement instruments; virtually all engineering departments in universities will have their own calibration facilities; and, of course, there is a plethora of calibration companies. You are free to select any of these, but you *must* have a calibration certificate for each instrument you intend to use.

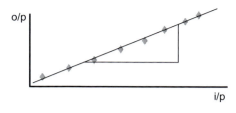

Figure 9.6
Sensitivity.

The main aim of all calibration activities is to obtain a graph of input versus output. This graph reveals a plethora of information.

9.3.3.1 Sensitivity
The usual protocol to follow is to vary the input to a device and measure the output. These values are then plotted on a calibration graph with input as the abscissa (horizontal) and the output on the ordinate (vertical).

The sensitivity is the gradient of the best-fit line through the points. This graph is easily obtained using a spreadsheet (such as Excel); but be careful to use the x-y scatter graph. Obtaining a best-fit line is also easy and the gradient is easily obtained. Using a spreadsheet or a data analysis package is by far the best method as the statistics are all done for you.

You are able to produce a nonlinear system, but in this case sensitivity will also be nonlinear. Sensitivity will then be an equation in the form of a polynomial, moving average, or other suitable function. These are, of course, far more difficult to deal with and, in general, are to be avoided.

9.3.3.2 Range
Most real devices produce a saturation curve. This has three distinct regions. At low levels of input the physical errors in the system (internal friction, etc.) make measurements unreliable and as a consequence the output is nonlinear (output is not proportional to input). At larger inputs the measurements, again, become unreliable as the device's limits of operation have been exceeded. Once again, the output is not proportional to input. In between there is, normally, a region where the device behaves itself and the output is proportional to input: the device is behaving linearly. The region of the input where this is true is called the *range* of the device (Figure 9.7).

Once again, spreadsheets and data analysis packages come to the rescue; range is determined with relative ease.

9.3.3.3 Repeatability
Repeatability is a measure of, well, whether the outputs are repeatable. To put this another way, for the same input do you always get the same output? Conducting the input–output experiment repeatedly and plotting all the points on a single graph will allow you to obtain this measure (Figure 9.8).

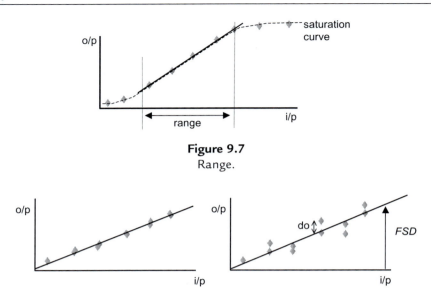

Figure 9.7
Range.

Figure 9.8
Repeatability: (a) good repeatability; (b) poor repeatability.

Repeatability is defined as the largest variation in output (indicated by the small arrow), *do*, as a percentage of full scale deflection. Full scale deflection (FSD) is the largest possible output over the range of the device. Hence

$$R_e = \frac{do}{FSD} \times 100\% \qquad (9.2)$$

9.3.3.4 Reproducibility
Reproducibility is similar to repeatability, but in this case others conduct the repeat experiments. The calibration is conducted in a similar manner, using similar protocols, but different people use the device in different situations (e.g., different hospitals) (Figure 9.9).

Reproducibility is defined similarly to repeatability:

$$R_o = \frac{do}{FSD} \times 100\% \qquad (9.3)$$

9.3.3.5 Resolution
Resolution is defined as the smallest change in input that creates a discernible change in output. It is akin to comparing two people walking down a road, one being very tall, and the other very short. Both cover the same distance, but the shorter chap does it in smaller steps. Hence if your device uses a rule to measure distance then its resolution is the smallest division.

Since we all went digital, resolution has become extremely important. Systems use analog to digital converters (ADCs) to change continuous analog signals into a simple stream of numbers. However the resolution of an ADC depends on the range of the input and the number of bits.

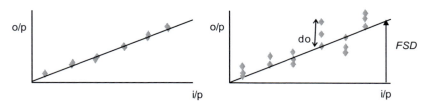

Figure 9.9
Reproducibility: (a) good reproducibility; (b) poor reproducibility

---

CASE STUDY 9.4

A 12-bit converter is used to measure a voltage of 0-1 V. The range of the ADC is 0-10 V. Determine the resolution and suggest an improvement.

The number of divisions of an ADC is given by 2^n, where n is the number of bits. Hence a 12-bit converter has 4096 divisions. Therefore the resolution of the ADC is

$$\Delta R = 1/4096 = 0.00024V$$

Inserting an amplifier of gain 10 between the signal and the ADC can improve the resolution. This now uses the full range of the ADC and changes the resolution to

$$\Delta R_k = \Delta R/k = 0.00024/10 = 0.000024V$$

Please note that modern televisions and cameras use the term *high definition* and relate this to high resolution. This is in fact wrong. TV sets with better definition almost certainly have low values of resolution!

I am always amazed how people have accepted digital information to be more accurate than analog. This is not the case; it is certainly easier to deal with but it is by no means more accurate.

9.3.3.6 Linearity
The last calibration term for us to consider is *linearity*. This is simply defined as the greatest deviation away from linearity as a proportion of FSD (Figure 9.10):

$$L = \frac{dl}{FSD} \times 100\% \tag{9.4}$$

9.3.3.7 Summary of Calibration
It cannot be stressed enough that calibration is important whatever your device. It is, by far, the best way to confirm that your device is performing as it should. You may not need all of the terms we have met and, equally, these are not all of the terms associated with calibration.

Consider the risk analysis associated with two dialysis machines where one has been calibrated and the other has not; we know exactly how much flow is generated for a particular

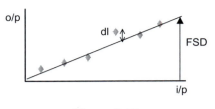

Figure 9.10
Linearity.

CASE STUDY 9.5

A pump has been selected for use in an infusion pump. The output of the pump is supposed to be proportional to applied voltage. The pump's characteristics were determined and plotted as a graph of output flow rate (ml/min) versus applied voltage (Figure 9.11). Determine any obvious calibration data.

The pump has an identifiable range, 1–5 V. To go the next stage – calibration – we must remove the outliers (below 1 V and above 5 V). We can then fit a straight line to the data using a "best-fit line" routine (in most spreadsheets this involves "fit trendline") (Figure 9.12).

From this single graph we determine that

$$\text{Range} = 1 - 5\,\text{V}$$
$$\text{Sensitivity} = 5.0025\,(\text{ml/min})/\text{V}$$

The maximum deviation of any point from the straight line is 0.1 ml/min and FSD is 25 ml/min; hence, using Equation (9.4), linearity is given by

$$L = \frac{0.1}{25} \times 100\% = 0.4\%$$

We are unable to determine any more data from this graph.

---

setting and there is a document attached to the machine to state when it was last calibrated (and when the next calibration is due); the second machine has no such providence. Which do you think is the most risky to use? Obvious, is it not?

9.3.4 Surface Evaluation

There are two reasons to conduct surface evaluation. The first is, primarily, a quality issue and is related to the confirmation of surface finish. The second is primarily due to the growth in the use of surface coatings; one such example being the coating of implants with hydroxyapatite. There are numerous methods for surface evaluation, but I do not intend to go into great depth. Their relative cost increases dramatically with the degree of magnification; but some are well within the "common man's" reach.

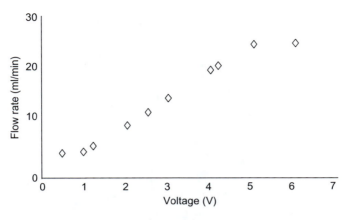

Figure 9.11
Data for Case Study 9.5.

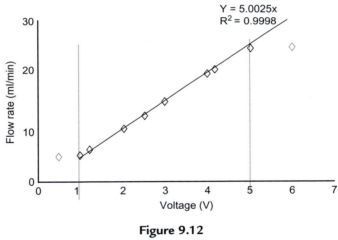

Figure 9.12
Calibration graph.

A word of warning: the costs increase inordinately with magnification. If you are going to pay for the higher magnifications then make sure you have specific criteria to measure against. Whatever the device, make sure that it has been calibrated, and check the calibration before use.

9.3.4.1 USB Microscopes

Although specifications state they can reach magnifications of X400, do not expect too much. They often sell for less than $50 and hence you get what you pay for. But for inspection of

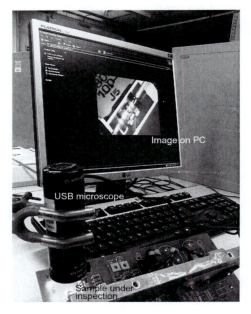

Figure 9.13
Typical USB microscope. *(Courtesy Staffordshire University)*

surface damage, crack propagation, etc., they are a very affordable device (Figure 9.13). All are able to take still images and some come with the ability to measure dimensions too.

9.3.4.2 Commercial Optical Microscopes

The price of these devices depends on the quality of the lenses and the magnification achieved. They are still affordable and many have the ability to take still images using a digital camera (Figure 9.14). These devices are often calibrated and (unlike USB microscopes) it is possible to take accurate measurements from the images.

9.3.4.3 Profile Projectors

These devices rely on the production of an accurate silhouette of a component. Most have digital measurement capacity and can magnify an image significantly. They are calibrated devices and can often measure to an accuracy of 1 μm (1/3000 of a human hair). They are not expensive (about $3000–5000) and are the mainstay of all commercial quality departments. Many have the capability for surface evaluation using surface lighting (but not all). However, these devices are best used for checking for wear of profiles after accelerated life tests (Figure 9.15).

9.3.4.4 Hardness Testing Machines

Often one of the results of accelerated life tests is work-hardening of the component. Hardness testing machines are invaluable for this type of surface evaluation. However, they are beyond

Figure 9.14
Typical optical microscope with digital capture. *(Courtesy Staffordshire University)*

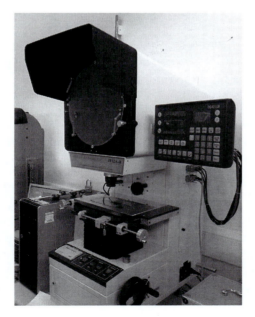

Figure 9.15
Typical profile projector with digital display. *(Courtesy Staffordshire University)*

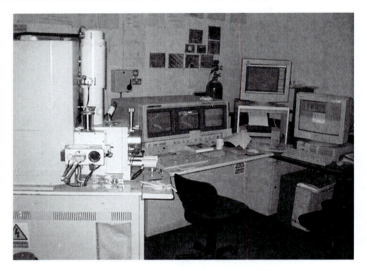

Figure 9.16
Typical scanning electron microscope. *(Courtesy Staffordshire University)*

the reach of most companies (often priced around $10,000) and need specialist training to use. Most engineering colleges and universities will have one or more in their materials department.

9.3.4.5 Scanning Electron Microscopes (SEM)

These are unaffordable for all but the largest of global companies, even the annual running costs would make most technical managers' eyes water. They have the ability to measure at sub-μm levels (magnifications of up to X500,000) and are the mainstay of virtually all surface evaluation studies (Figure 9.16). However, because of the detail and amount of information supplied, some studies can suffer from information overload.[1]

9.3.4.6 Atomic Force Microscopy (AFM)

This is another very specialized device that only few commercial industries would own, but many universities specializing in nanotechnology would have one. The device enables visualization of a surface to the nanometer (1×10^{-9} m) scale. They can also be configured to measure electric potential.

9.3.4.7 Beam Profile Reflectometry (BPR)

Beam profile reflectometry (BPR) uses a low-power, focused laser beam to analyze the surface and return information about the coating's thickness, refractive index (which is closely related to density and composition), and even strain or other structural anisotropy. It can even cope with surfaces that have complex shapes and/or high local curvature. The technology, which originated in the semiconductor industry, has been adapted for use on

---

[1] Information overload is when too much data is presented such that it actually confuses those that are nontechnical. Sometimes this is unintentional. Unfortunately it is sometimes intentional in order to suggest scientific rigor when there is, in fact, none.

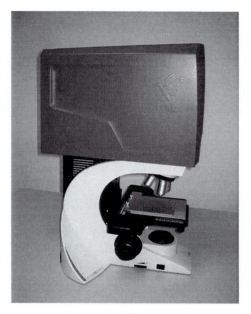

Figure 9.17
Desktop BMR device. *(Courtesy Nightingale-EOS Ltd)*

medical devices by Nightingale-EOS Ltd and forms the basis of their n-Gauge™ coating metrology tool. Transparent coatings and stand-alone membranes in the range ~0.1 µm to 150 µm can be measured in this way (Figure 9.17).

9.4 Computer-Based Evaluation

The power of modern computers and associated software is at a level where many aspects of a design can be evaluated virtually while it is self-evident that only final physical evaluation gives a true picture, it is also true that the final evaluation need only be that, final. For example, it is stated that the latest Airbus was designed and tested wholly on a computer. However, how many people would believe it could fly until it actually did (Figure 9.18)? Computer-based evaluation can be divided into the categories listed in Sections 9.4.1 to 9.4.5 below.

9.4.1 Animation

Many dynamic aspects of a device, such as mechanism movement, can be tested using the built-in animation facilities within CAD packages. Animation is an extremely useful tool for checking assembly and disassembly protocols.

9.4.2 Dynamic Simulation

Most modern CAD packages come with built-in motion simulation. This is powerful enough to detect collisions and also to predict forces within use.

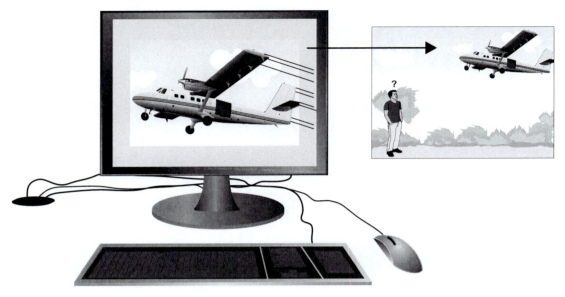

Figure 9.18
Computer-based evaluation dilemma.

9.4.3 Finite Element Analysis

FEA is now a fully accepted analysis system. The predictions it makes for stresses, strains, etc. are wholly acceptable as predictions of "real life."

9.4.4 Computational Fluid Dynamics

As with FEA, CFD is an accepted system for evaluating the performance of a device – provided you follow the GIGO principle.

9.4.5 Caveat

Any computer-based evaluation package you use will come with a caveat stating "the use of this software does not replace physical evaluation." While this is true, there are certain circumstances where physical evaluation and simulation coincide. You will need guidance by qualified, chartered/licensed engineers to make this judgment.

9.5 Value to "Healthcare" Analysis

At some point you will need to prove your device to the purchasing body, irrespective of the support given by clinical staff. This will always revolve around the "value" of the device.

- Will it be cheaper?
- Will it save money in the short term?
- Does it have economic benefits in the long term?

Unfortunately, innovative devices will hardly ever be cheaper. They will, almost certainly, be more expensive than the alternative. Hence you will need to provide evidence of cost benefit, not just health outcome. This falls under the banner of *health economics*.

Health economics is a discipline unto itself. It is fraught with typical accountancy jargon and quasi-scientific principles. However there are certain steps the average designer can undertake without resorting to paying a health economist.

9.5.1 Distinct Health Benefits

Do you understand the real health benefits of your device? Have you explored how your new device will affect the clinical community? How does the device benefit the end-user? Sometimes benefits are direct, sometimes indirect. Table 9.6 attempts to illustrate these.

You must also consider to whom the benefit applies. For example a patient under general anesthesia will have better morbidity outcomes, a clear benefit for the patient. However this also impacts on the hospital's ratings (where deaths per procedure are normally counted); it also reduces the cost of the procedure since less anesthesia has been used. This then has a knock-on effect as the anesthetist is free to move onto the next procedure with greater ease.

From this example you can see that a simple benefit has a massive knock-on effect that can snowball into a large overall benefit. Table 9.7 lists some benefits you may wish to consider. This is by no means complete but it gives you an idea. Remember that your list of people who benefit can expand as far as you wish, and you can drill down in detail (e.g., central sterilizing department, scrub nurses, etc.)

Table 9.8 takes Table 9.7 and uses it to start to illustrate how the benefits may be exhibited. It is important to note that the benefits need not be direct; they can be knock-on benefits.

Table 9.6: Examples of Direct Benefits Generating Further Indirect Benefits

Location	Direct	Indirect
Operating theater	Shorter operating time	— More ops per day — Shorter GA duration (morbidity)
Operating theater	More predictable operation duration	— Improved planning — Higher utilization
Clinic	Faster clinical assessment	— More patients per clinic — More time for clinician
Sterilization unit	Parts easy to identify	— Less chance of lost items — Less chance of operation cancellation — Better productivity

Table 9.7: Example Benefits

Benefit	Patient	Clinician	Healthcare Provider	Society
Treatment duration	✓	✓	✓	
Time to work	✓			✓
Less referrals	✓	✓	✓	
Less medication	✓	✓	✓	✓
Less time in hospital	✓	✓	✓	✓
Less time under GA	✓		✓	
Faster detection	✓			✓
Reduces clinic time		✓	✓	
Easier to assemble prior to use		✓	✓	
Easier to use		✓	✓	

Table 9.8: Example Expansion of Benefits

To whom	Direct	Indirect
Patient	Shorter duration	Reduced trauma Reduced exposure to GA
Patient	Less invasive	Reduced trauma Smaller scars Shorter time-to-work
Clinician	Easier to use	Reduced times More quality time with patient Greater user satisfaction
Clinician	Greater accuracy	Reduced times Less trauma to patient Greater user satisfaction Better clinical outcomes
Healthcare provider	Easier to use	Lower costs Higher throughput Less revision Lower litigation
Healthcare provider	Lower infection rate	Less antibiotics Lower referral rate Lower cost
Patient	Lower infection rate	Less antibiotics Less trips to clinic Shorter time-to-work

You must identify all benefits, and then quantify them in order to persuade the hospital/healthcare provider to purchase your device. You can bet your shirt that they will want numbers, and those numbers must be prefixed with a currency symbol!

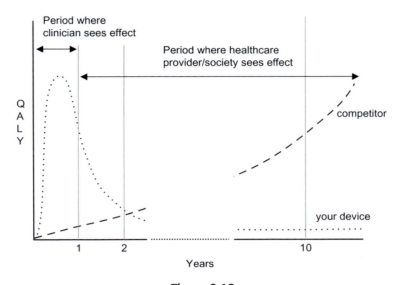

Figure 9.19
Example of a QALY graph demonstrating benefits over a long period.

If you find the development of Table 9.8 daunting, try using the radial thinking technique (and the other creativity techniques) illustrated earlier.

It is worth mentioning quality of life. Some healthcare providers use QALY as a measure (Quality Adjusted Life Years – most now write QUALY). It is a very simple concept. It is a measure that takes into account a small benefit to quality of life, but one which is enduring as opposed to a quick fix remedy that has limited life. For example your device may have an immediate requirement for antibiotics and painkillers, but this requirement lasts a few weeks. This is compared with a competitor device that has no need for antibiotics but which has the patient taking painkillers for the rest of their life. A QALY measure over the first year would probably put your device well down the table of "good devices." However, measuring QALY over the whole lifespan of the patient puts your device out in front. Figure 9.19 attempts to illustrate this in graphical form. The graph also shows that portions of the graph have greater influence on different end-users. For example a clinician may only see the patient during the early treatment and then have a follow up after 12 months. Beyond that their interaction with the patient may be zero. However, long-term costs for treatment of pain, etc., are borne by society and the healthcare provider hence they are very interested in the aftereffects.

9.5.2 Stating Clinical Benefits

While you may have assessed your device's benefits and have some idea of what it can achieve, until this is proven you are not allowed to state its benefits in any literature or marketing material. Hence there is a need for an evaluation by a clinician. There is little doubt that the power of that message is stronger if the clinician has no link with your company. If your device is CE marked then any clinician can use it to determine the significance of its performance in real life. Hence, for example, if you think your device can be assembled in half the time of your competitor's device then there is no point you stating this. If a clinician demonstrates this and publishes it independently then your message is strong. Also it is *not* your study hence you need not worry about *ethical* issues. Any clinician is allowed to perform their own studies using currently CE marked devices under their own ethical procedures. However you have to wait for someone to think of doing this, on their own.

If, however, you cannot wait for this clinician to turn up, you will have to pay for the service. It is almost certain that you will be unable to do this without outside help. Now you have entered the ethical circle and you are paying the clinician or study team to undertake this aspect for you. Most university hospitals, indeed most universities, offer this service. However, it is *not* a clinical trial…that is something completely different. This is an evaluation to identify any statistically significant differences or benefits. This is a clinically led study; it is *not* a clinical trial, nor is it a clinical evaluation. Do not get them confused as the confusion could bring the metaphorically long legislative arm of the law down onto your metaphorically tender shoulder.

I'm afraid it's time to mention statistics. We shall look at hypothesis testing in more detail when we examine clinical trials. However, it is worth noting that the whole of the clinical effectiveness world works on hypothesis testing and values of p. The magic number, often quoted, is $p<0.05$ – what does this mean? Later we shall see that the statistics will almost certainly be examining two groups and the difference between them. We look for a small value of p as this demonstrates that the two groups *are* different, and we hope that the only difference is your device!

Golden rule: do not state *clinical* benefits unless you have the evidence to do so.

9.6 Clinical Studies and Clinical Trials

The distinction between a clinical study and a clinical trial is very clear. If you are unable to put your device onto the market without first performing a clinical study proving efficacy and safety then this is called a *clinical trial*. If your device can be CE marked without a clinical study, then it is *not* a clinical trial and does not require the same amount of legislative rigor. In essence the actual tasks undertaken are the same, it is the paperwork and pretrial licenses required that make them different. Whatever, or whichever, you are attempting you should be working to ISO 14155:2011 (ISO, 2011). Before we delve into ISO 14155 let us examine some critical terms.

9.6.1 The Hypothesis

Why are clinical studies and clinical trials (collectively known as *clinical investigations*) effectively the same? Very simply, they both aim to test a hypothesis. "*What is a hypothesis?*" I hear you ask. It is the important part of any clinical study, be it a trial or not.

"A **hypothesis** *(from Greek) consists either of a suggested explanation for a phenomenon or of a reasoned proposal suggesting a possible correlation between multiple phenomena. The term derives from the Greek, hypotithenai meaning 'to put under' or 'to suppose.'"*

(Wikipedia, 2011)

"An unproved theory. Proposition, etc."

(Collins, 2009)

For medical devices the *hypothesis* is the *clinical benefit* that your device produces and the trial is trying to prove it is *real*. The clinical benefit need not be positive, it can be neutral or even negative. But the wording is quite specific and you, as the designer, need to have a good idea of what it should be. To go back to our design methodology, you are in fact designing your study and hence use the same methodology that we have seen throughout. Therefore, for the study the hypothesis is your statement of need. For example Table 9.9 illustrates some benefits and a typical hypothesis.

Note that the hypothesis is a statement that is either *true* or *false* (*null*). Hopefully you are trying to prove that your hypothesis is true – therefore enabling you to make a statement about the clinical benefit.

Where a trial deviates from a study is in the examination of side effects; that is, a thing going awry. If you have conducted you design correctly failures should be avoided. However, you can never escape the unexpected long-term effects that will only occur after long-term use. In essence, and before you can release your product, you must have tested the hypothesis "This device is safe to use."

If you cannot test this hypothesis and obtain an emphatic "true" result then you need a clinical trial to test it. If all of the precedents, design work, and evaluations you have undertaken make the answer a "true," then a clinical trial is probably unnecessary. Unfortunately there is an exception to this rule; under current EC rules *"All devices of classification class III require clinical trials"* (EC, 2010).

Table 9.9: Clinical Benefit Hypotheses

	Example Clinical Benefit	**Example Hypothesis to Be Tested**
Positive	The rate of non-intervention has increased	The use of device A for the treatment of B results in an increase in the rate of non-intervention when compared with traditional methods.
Neutral	The infection rate is no worse compared with others	The use of device C for the treatment of D results in infection rates that are no different than those observed when device(s) E (F and G) is (are) used.
Negative	The amount of antibiotics taken is reduced	When treating X, the amount of antibiotics used is lower when device Y is used compared with common practice.

And that is irrespective of precedents, evaluations, and design studies and it applies in a similar way in all markets around the world. This ruling is also starting to be applicable to all implants and to all new software.

Is it possible to test more than one hypothesis at a time? Clearly, if you are testing the safety of a device you can also evaluate clinical benefits as often the measure is simply, do the benefits outweigh the risk? However for clinical studies it is difficult, if not impossible, to conduct a proper study if more than one hypothesis is being tested. So the simple answer is no; try to avoid multi-hypothesis studies.

The study is a part of your design process so it needs documenting. Hence it is preferable to have a pro forma that states the hypothesis you are intending to test, and gives any further information/aims that may be required. Figure 9.20 illustrates a typical pro forma that bares a remarkable resemblance to a statement of need.

9.6.2 Investigation Specification

You should not be surprised to see that the starting point is in fact none other than a PDS for a clinical study. You should be designing your study using the same methodology as the design of the device, and the starting point is the PDS. A word of advice: always include a statistician (or at least someone with a working knowledge of statistics) at this phase. Too many studies have collected results from a poorly designed study (from a statistics perspective) only to result in no meaningful results being produced. Hence one should always start with a PDS for the study. You can use the same pro forma as shown in Chapter 5, but your sections will be different, as follows:

1. **Target population**: Who is the beneficiary? This may be an age group, a gender group, maybe even an ethnic group. The beneficiary may not be a patient; it could be a member of hospital staff, such as a nurse or surgeon.
2. **Regulatory and statutory**: If your study includes patients or human subjects it is required to abide by the Helsinki Declaration[2] (now in its sixth revision); we will discuss this further later. If your device is not cleared for market in the USA nor has a CE mark then you will need to seek approval from the appropriate regulatory body. Does your study fall under any specific national or international standards? Do you need to meet any standards before the study can start? If you are going to capture patient data will you be bound by data protection laws or freedom of information requirements?

---

[2] The Helsinki Declaration is concerned with the ethical issues related to experimentation on human beings.

Medical DeviceCo Inc.

Clinical Investigation Hypothesis

Project Number

Product Title

Description of Device

Hypothesis

Further Information/Aims

Approved/Not Approved

Signed

Date

SoH.doc   version 1.0

Approved by:
Date: 17.5.2010

Figure 9.20
Typical statement of hypothesis pro forma.

3. **Exclusion/inclusion criteria**: Some now refer to inclusion criteria in a politically correct manner – it is not. Both relate to the selection of subjects for a study. The inclusion criteria select those who must be in the study; for example they must have a broken leg. Those to be included in the study should be those whom your device is intended to treat. The exclusion criteria, effectively, removes subjects (from the inclusion set) from the study – normally any one from a list. Typically this may be smoking or drug dependency, but it could be children and/or the aged.

4. **Data requirements**: You should try to imagine all the data you may need and list them. It is actually better to collect as much information as possible. However, some data may be

---

Medical DeviceCo Inc.

Clinical Study Spcification

Originator	Date

Project Number   Version:

Product Title

Hypothesis

1. Target population:

2. Regulatory and statutory:

3. Exclusive criteria:

4. Data requirements:

5. Suggest study type:

Approved/Not Approved

Signed

Date

CSS.doc   version 1.0   Approved by:
                                            Date: 17.5.2010

Figure 9.21
Example clinical study specification pro forma.

difficult and expensive to collect (e.g., MRI scans) so this has to be thought through with great care so as not to miss something. This section should also estimate the number of subjects required. This is where you need advice from a statistician.

9.6.2.1 Study Types

There are two main study disciplines: prospective and retrospective. A *prospective* study starts with a plan and works from a blank page – it is forward looking and is controlled. A prospective study is the "gold standard" and your marketing staff will love you forever if you obtain marketing data using this methodology.

A *retrospective* study looks backward in time, has no plan, and has no real hypothesis; its sole aim is to look backwards and identify any trends or averages. In real life retrospective studies are frowned upon as they have not been controlled…however in some cases this is all a company has so they have to make do with it. However even retrospective studies are bound by the usual ethical protocols as one is accessing private data for use other than clinical treatment. Commonly, registrars (interns) tend to do this type of study as a part of their training and it is normal to go to clinical conferences and be bombarded with presentations, from newly qualified surgeons, that are clearly retrospective in nature: take them with a pinch of salt as the data is tainted by the lack of control.

In common practice, there are three main types of prospective study:

- **Open:** In an open study one is not comparing one device with another; one is only examining one device. This may be to obtain an average time to set up, or an assessment of usability. But the data is not for comparison (except against historical norms, which is not a good idea as there is no control). This is the sort of study one sees for toothpaste, etc. where a statistic is stated: *"84% of users said it was wonderful and would recommend it to a friend,"* or *"88/90 patients returned to work."* The trouble with this type of study is that for all we know 99% of existing users think the same for the existing device, or 90/90 patients return to work with the existing device…I think you can see the problem. You will need to persuade clinicians, procurement, medical staff, etc. that yours is the best device to buy; they will always come back with a question concerning comparison.
- **Blind:** A blind study is designed to take out the influence of the subject. This is normally achieved by splitting the study into two subject groups; the first group is treated using your new device, the second group (called the *control group*, see Table 9.10) is treated using normal practice. It is important that the second group is treated using normal practice for two reasons: firstly you can compare your results directly; and secondly your control group is not being disadvantaged – they are still receiving the best possible treatment. In no circumstances does this mean the use of a placebo.[3]

---

[3] A placebo device is one that has no effect at all. Would you be willing to have this treatment?

Table 9.10: Study Control Group Types (MHRA, 2011)

Control Group Type	Description	Comment
Concurrent	Both groups, and all within, are treated by the same person.	Gives the study good control but causes an issue of transferability: Are the results only applicable to the one clinician? Get over this by using multicenter studies.
Passive-concurrent	The control group does not receive treatment from the same person as the active group.	Less controlled but any differences may be due to the different clinician.
Self-control	The subject acts as the control and crosses from control to active group under their own volition but following a transition protocol.	Only useful for devices that treat long-term or chronic conditions. Subjects cannot suddenly perform an operative procedure on themselves.
Historical	A retrospective study against a group separated by time.	As stated previously, retrospective studies are frowned upon.

This is not a route to be advised; even though all statistical guides say they can be used, most clinicians do not accept the "do nothing" approach. However, no individual subject knows which group they are in, so you have to do your best to make sure that they cannot identify which group they are in! In this way the subject is effectively "blind" and has no influence on the outcome. Randomization is an important aspect of a blind trial; subjects are put into groups at random and they are *not selected.*

- **Double Blind:** A double blind study removes the influence of the investigator as well. In this study nobody knows who is in which group. This is usually achieved by having an external statistician who controls the subjects' ID and their grouping but does not share this information. Only when they start to examine the data does the grouping become apparent. Once again random selection is important. A double blind trial is the gold standard in clinical investigations.

All of the categories above can be *controlled* or *uncontrolled*. Having a control group does not control the study; it is only control by name, not control by design.

- **Uncontrolled**: The investigator lets everything happen by chance with no consideration of the effects of any variables.
- **Controlled**: One or more variables[4] are controlled. This is normally controlled using the exclusion/inclusion criteria. However you may wish to control factors that are not directly subject related but are directly related to the operation of your device and its alternative.

---

[4]A variable in this context is usually subject based, for example body weight, age, gender.

I suspect you are beginning to understand why the planning of a study is a professional job. However, this does not give you the excuse to ignore your part in the planning stage. You need to write a proper brief otherwise the "professional" can lead you a merry dance and you will pay a lot of money for no worthwhile result.

9.6.3 Relationship with Ethical Committees

ISO 14155 (and just about every other guide you care to mention) will stipulate that all studies on human subjects require approval by an ethical committee. This is where your links with a local university or teaching hospital come into action. It is highly unlikely that you will have your own ethical committee, but teaching hospitals and universities do. It is the ethical committee that approves the study, and to meet the requirements of ISO 14155 you need this approval documented. Unfortunately this will not be free, but most of these establishments are looking for research projects for their professors and research staff and so long as you are willing to allow publication (by an independent body – your marketing department will love this) of the results, some form of financial arrangement is always possible.

It is important to note that not all hospital and university ethical committees work to ISO 14155. This is because their ethical committees look at more disciplines than simply medical devices. To ensure your study meets the FDA and EC requirements you must ensure the respective ethical committee understands that your study must meet the requirements of ISO 14155 even if it exceeds their individual requirements.

9.6.3.1 Informed Consent

As a part of the ethical approval process you and your investigation team will need to produce a document that enables your subjects to give informed consent. This document is best written by those trained in their production. However as the "sponsor" you will need to ensure that it is done, approved, and enacted. It is unlikely that any teaching hospital or university will want to lose its ability to give ethical approval and shortcut the process, but you still need to be wary of rogue investigators (especially in nation states where human rights may not be as well established as we would expect).

9.6.4 Relationship with Regulatory Bodies

If your product is already CE marked (or the FDA equivalent) then it is probably not necessary for you to notify your regulatory body. However, if you are taking the device outside of its agreed indication for use then you may well need to. If in doubt, contact them directly and discuss what you are going to do with them. As I have stated earlier they are not ogres, they will help – if they can.

If your product does not have a CE mark or 510k and this study is a part of your clinical evaluation prior to approval then you *must* notify the regulatory bodies and obtain formal approval for the study to go ahead as this is now a clinical trial. This process will go hand-in-hand with the ethical approval as nothing can go forward until both have been formally approved. Both the FDA and EC regulatory bodies have guidelines and procedures in place.

ISO 14155 makes it *your* responsibility as the *sponsor* that the study is conducted correctly and that all relevant documentation is in place. You can delegate the responsibility of design and action to a professional, but in the end *you* or *your company* have overall responsibility.

9.6.5 ISO 14155 and EC–FDA Guidelines

ISO 14155 is the overarching standard for clinical investigations of human subjects. If you use this standard as the basis for your clinical study you will be meeting every requirement of every regulatory body and, at the same time, ensuring you fulfill the requirements of the Helsinki declaration. The EC has a clinical evaluation guideline, MEDDEV 2.7.1 (EC, 2009), in which clinical studies are referred to. The FDA guide is in the design control guide (FDA, 1997). However, working to the ISO will meet their respective requirements. Both have detailed guidance as illustrated by Table 9.11.

It is obvious that as technology improves the guidelines change and, as you can see in Table 9.11, they are constantly being updated. You should, therefore, keep a keen eye on these guidelines as you do your standards portfolio. It is interesting to note that the FDA guidelines do not refer directly to ISO 14155; however a scan of the guidelines makes it clear that meeting ISO 14155 means you have, effectively, FDA requirements too. However, as with all other regulatory statements, check first!

Table 9.11: FDA and MHRA Guidance Documents for Clinical Investigations

Title	Body	Published or Draft
1: Guidance for manufacturers on clinical investigations to be carried out in the UK	MHRA	Published
3: Information for clinical investigators	MHRA	Published
4: Pre-clinical assessment guidance for assessors	MHRA	Published
17: Guidance notes for manufacturers on statistical considerations for clinical investigations of medical devices	MHRA	Published
Design considerations for pivotal clinical investigations for medical devices	FDA	Draft
Investigational device exemption (IDE) for early feasibility medical device clinical studies, including first in human (FIH) studies	FDA	Draft
The 510(k) program: evaluating substantial equivalence in pre-market notifications [510(k)]	FDA	Draft
Statistical guidance for clinical trials of non-diagnostic medical devices	FDA	Published
Guidance for the use of Bayesian statistics in medical device clinical trials	FDA	Published
FDA decisions for investigational device exemption clinical investigations	FDA	Draft

9.6.6 Analysis of Data

Without exception, the analysis of the data obtained from your study will entail some statistical analysis. That is why your study must be designed to meet statistical considerations at the start. For most scientists and engineers the thought of performing a statistical analysis is not daunting. However, for many manufacturers statistical analysis may be as alien as life on Mars. It is beyond the scope of this text to teach you statistics, but there are a number of tools the average computer literate person can attempt to use. The first is correlation and the second is student t-tests; both are openly available in spreadsheet packages such as Microsoft Excel®. If you want to read more then there is a plethora of textbooks on the subject. Look for ones that are targeted at clinical studies or clinicians as these are directly relevant.

9.6.6.1 Outliers and Missing Values

In any trial there will always be exceptions to the rule. Some subjects will just "disappear," often quite logically, from the study. These are called *missing values* and can legitimately be excluded from any analysis, but you must state so in the analysis report.

Outliers are subjects whose results are way off the norm. This is normally due to some congenital, physical, or historical reason. For example, one of your subjects may have been a habitual smoker and only recently stopped. Hence when you asked "Are you a smoker?" they reply "no"; as a consequence their result may well differ greatly from everyone else who replied "no" to being a smoker. This is a good reason to revisit your exclusion criteria; it is a valid reason for excluding the data from the analysis. Another valid reason may be to exclude all subjects with healing times 2× the average value; but this one is subjective and should be stated in the study plan. It is *not* acceptable to simply remove data because it distorts your averages: this is called *fixing the data*.

9.6.6.2 Correlation

One of the methods for confirming that the output (or the result of your study) is due to your intervention is to examine the correlation. Correlation, effectively, analyzes the data accepting that there will be a naturally occurring scatter in the data and that you will need to see that if you change A then B follows. The best way to illustrate this is by example. Suppose you had a study examining the effect of the length of time your device was used on the pain relief obtained. One may have a table of data such as illustrated in Table 9.12.

A first look at the data suggests that the device causes a reduction in pain. If we plot a graph of duration versus difference in pain score we can fit a straight line to the data. If we use a spreadsheet to do this then we can request to show the line equation and the R^2 value; this is the correlation coefficient (Figure 9.22).

The straight line suggests that as duration increases the reduction in pain increases. However the correlation coefficient $R^2=0.08$. Table 9.13 indicates typical acceptance values of R^2 for $p=0.05$.

Table 9.12: Example Pain Scores Following a Study

Sample	Duration (Minutes)	Pain Score		
		Start	End	Difference
1	9	10	8	−2
2	4	5	4	−1
3	8	9	9	0
4	8	9	9	0
5	7	8	7	−1
6	2	3	2	−1
7	7	8	6	−2
8	3	4	3	−1
9	5	6	4	−2
10	7	8	6	−2
11	6	7	6	−1
12	1	2	2	0
13	4	5	5	0
14	9	10	8	−2
15	6	7	6	−1
16	6	7	5	−2
17	8	9	7	−2
18	7	8	6	−2
19	1	2	1	−1
20	1	2	1	−1
21	7	8	7	−1
22	8	9	9	0
23	2	3	2	−1

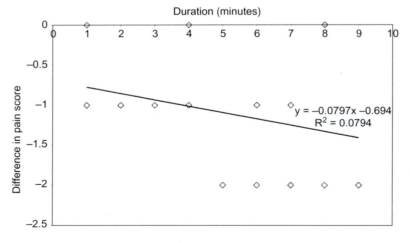

Figure 9.22
Example correlation graph.

Table 9.13: Typical Correlation Coefficients for *p*=0.05

No. of Points	R^2 Equal or Greater Than
5	0.88
10	0.63
15	0.51
20	0.44
25	0.4
30	0.36
50	0.28
100	0.2
1000	0.06

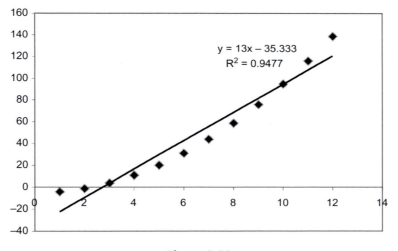

Figure 9.23
Illustration of a good correlation result but clearly a nonlinear graph.

Figure 9.22 yields an R^2 of 0.08; much less than the 0.4 required. Hence there is no correlation between duration and pain score. Take care: always check the graph as statistics can lie because you may be trying to fit a straight line to a nonlinear data set (as illustrated in Figure 9.23).

9.6.6.3 Averages and Confidence Limits
An average value is arguably the most commonly quoted statistic in the world. It is, however, the most misused. Quoting a single value is useless, as Figure 9.24 illustrates. All three data sets have the same average, but the spread of the data is wildly different.

This spread (or scatter) comes from naturally occurring variations (scientists and engineers spend their whole lives trying to reduce this scatter). Some people, erroneously, describe the

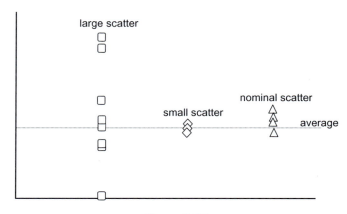

Figure 9.24
Demonstration of equal average but wildly variant scatter.

spread of data by giving the average and the range of the data. It is better to give an average and the confidence limits. These are normally the 95% confidence limits (or levels); and it literally means that you are 95% confident that the average lies within this range.

If you have 20 or more samples the equation for the confidence limits is

$$95\% \text{ limits} \approx \bar{x} \pm 2\frac{\sigma}{\sqrt{N}} \qquad (9.5)$$

where σ is the standard deviation and N is the number of samples. You should note that the constant value, 2, is rounded up. For an infinite number of data points it is 1.96, for 20 points it is 2.09.

Most spreadsheets have automatic functions for both average and standard deviation calculations (as do most calculators), so it is not a difficult task to produce an equivalent Table 9.14 for any data set. Using data set 1 as the example, this would be cited as

$$\text{Average} = 9.1 \pm 0.137(95\% \text{ conf limits})$$

9.6.6.4 The Student t-test

The t-test (as it is known) is the first port of call for most investigators trying to ascertain whether there is any difference between two (or more) groups. Describing the mathematical basis behind the process is beyond the scope of this text, but it is a very powerful tool that, again, most spreadsheet packages have as an built-in function. The main aim of the t-test is to test a hypothesis. As we saw earlier, the hypothesis is the basis for the whole investigation, so you can imagine how important this first step into a statistical analysis has become.

To demonstrate how the analysis is performed the data in Table 9.14 will be used, and we shall compare data set 1 with data set 2 (i.e., the one with the least error with the one with the

Table 9.14: Data Sets for Figure 9.24

Data	Set 1	Set 2	Set 3
	9.00	21.65	10.69
	9.01	7.05	9.01
	9.02	9.86	10.71
	8.99	7.43	8.99
	8.98	10.63	10.67
	9.00	13.18	9.00
	9.00	0.17	10.69
	9.00	0.15	9.00
	8.99	0.41	0.20
	9.99	20.24	11.86
	9.00	21.65	10.69
	9.01	7.05	9.01
	9.02	9.86	10.71
	8.99	7.43	8.99
	8.98	10.63	10.67
	9.00	13.18	9.00
	9.00	0.17	10.69
	9.00	0.15	9.00
	8.99	0.41	0.20
	9.99	20.24	11.86
No. of points	20	20	20
Average	9.1	9.1	9.1
Std Deviation	0.305	7.565	3.195
Error (Eq. 9.5)	0.137	3.383	1.429
MAX	9.40	16.64	12.28
MIN	8.79	1.51	5.89

most error). One would normally assume that the difference between the average of the two data sets is zero (i.e., they are the same). One then needs to select some choices, for example equal or unequal variances? If they are the same then the variance should be the same, but to perform both calculations is easy so it is possible to present both. The final selection is two-tailed or one-tailed. This means is the average likely to be lower only or higher only (one-tailed), or could it be both (two-tailed)? The magic item we are looking for is the *p*-value. If the groups are the same (and we are using 95% confidence) then $p > 0.05$; if the groups are different then $p <= 0.05$. The more dissimilar they are the smaller the value of *p*.

Using Microsoft Excel® and using the t-test option in data analysis, we have already seen that the standard deviations are different, hence the variances are unequal. Performing this analysis on data sets 1 and 2 yields the results shown in Table 9.15.

As can be seen, *p* for the one-tailed assumption is 0.49; clearly the groups are identical. Hence there is no *significant difference* between the two groups. Table 9.16 analyzes a different data set to illustrate significant difference.

Table 9.15: Results of a t-test Assuming Unequal Variances

	Set 1	Set 2
Mean	9.10	9.08
Variance	0.09	57.22
Observations	20.00	20.00
Hypothesized mean difference	0.00	
df	19.00	
t stat	0.01	
p(T<=t) one-tailed	0.49	
t critical one-tailed	1.73	
p(T<=t) two-tailed	0.99	
t critical two-tailed	2.09	

Table 9.16 illustrates that the value of p is 0.01 for one-tailed. The groups are significantly different for a one-tailed test. The value of p is always double for a two-tailed test hence $p=0.02$. This is still $<<0.05$, hence the groups are significantly different. Why give both? It is down to you to select which you use. If you are absolutely sure your average can only move in one direction, either higher or lower but *not* both, then a one-tailed test works. If you are not sure which way it can move then use the two-tailed test. In most circumstances we are looking for improvements, hence it would be a one-tailed test. If you have more than one group you can do the same analysis for each group in turn.

9.6.6.5 Multivariant Analysis

In many clinical studies there is more than one variable that changes between subjects; this can be body weight, hair color, date of birth, phase of the moon when injured, etc. If you are to take these into account you need to perform a multivariant analysis. You can do this by redefining your groups into these subsets. But the most efficient way is to find a statistician to perform the analysis for you.

9.7 Literature Review

Since the recent changes in the Medical Devices Directive, the FDA and EC rules about performing a review of literature have become much more similar. The form of presentation may be different, but they are essentially the same:

- *Substantial equivalence*: if you are to minimize the amount of testing described earlier in this chapter then a proof of substantial equivalence is of paramount importance. In this part of the review you need to demonstrate that your device has, in some form, been in existence before. You then need to show how your device is equivalent.
- *Known issues*: this is predominantly a review of recalls and notifications to either the FDA or to any of the EC regulatory bodies. The aim here is to identify any recurring,

Table 9.16: t-test for Data Sets That Are *Significantly Different*

(a) Data Sets	
Set 1	**Set 2**
9.00	41
9.01	13
9.02	19
8.99	14
8.98	20
9.00	25
9.00	0
9.00	0
8.99	1
9.99	38
9.00	41
9.01	13
9.02	19
8.99	14
8.98	20
9.00	25
9.00	0
9.00	0
8.99	1
9.99	38

(b) t-test result		
Two-Sample t-test Assuming Unequal Variances		
	Set 1	**Set 2**
Mean	9.10	17.24
Variance	0.09	206.58
Observations	20.00	20.00
Hypothesized mean difference	0.00	
df	19.00	
t stat	−2.53	
$p(T<=t)$ one-tailed	0.01	
t critical one-tailed	1.73	
$p(T<=t)$ two-tailed	0.02	
t critical two-tailed	2.09	

known issues in devices similar to yours and then to show that you have either designed them out or accommodated them. Note this part feeds directly to the overall risk analysis.

- *Clinical/scientific literature*: predominantly to ensure that your device is utilizing best practices and even better if the literature shows that your device (independent of you) is the best.

9.7.1 Conducting the Review

Referring to MEDDEV 2.7.1 rev 3 and modifying terminology to suit the FDA results in a literature flow chart.

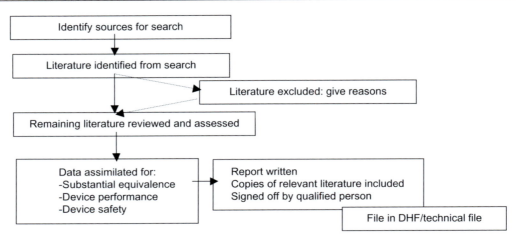

Figure 9.25
Literature review methodology *(adapted from* EC, 2009).

Figure 9.25 illustrates that potential literature needs to be found, selected, and analyzed. Potential sources for information are varied. There are several learned publication search engines for scientific literature:

–   Google Scholar: http://scholar.google.com
–   Medline: http://www.ncbi.nlm.nih.gov/pubmed/
–   OVID: http://ovidsp.ovid.com/
–   COCHRANE register of clinical studies: http://summaries.cochrane.org/

You will be able to see the abstract of the paper but unfortunately you will need to have an account with the relevant publisher to obtain a full paper. However, if you work with a local university they may have access via an educational license. In most cases the abstract is good enough to make some selection of the papers you really want access to.

Obviously standards are an important port of call, but registers of notices and recalls are equally valid (see Table 9.17).

Each of the search engines will require keywords for the search. You need to think of these carefully; also, you will have to write them down as a record of your search.

For substantial equivalence the 510(k) database is of great value.

9.7.2 Format for Literature Review

As with all we have seen so far the best way to approach this is to have a standard pro forma. Figure 9.26 is an example of a pro forma that would meet MEDDEV 2.7.1 rev 3, but also provides the evidence for any FDA application.

Table 9.17: Example Medical Device Recalls and Incidents Databases

IRIS – Australia's medical incidence database MAUDE – FDA database of manufacturers' experiences Medical device databases (FDA)	http://www.tga.gov.au/safety/problem-device-iris.htm http://www.accessdata.fda.gov/scripts/cdrh/cfdocs/cfmaude/search.cfm http://www.fda.gov/MedicalDevices/DeviceRegulationandGuidance/Databases/default.htm

Clinical Evaluation Literature review

Product Details	Part Number
	Title

Evaluation details	Name
	Date

Summary:

Approved by:
Signature   Date:

1. Methodology:
2. Outputs:
3. Analysis:
      Substantial Equivalence
      Device performance
      Device Safety

Figure 9.26
A typical literature review pro forma and sections.

Methodology: This section describes how the literature search was conducted and over what period it was undertaken. It should contain the names of the search engines used and the associated keywords and criteria. It should state the criteria for exclusion or the justification for inclusion of any identified sources into the main review. It is useful to attach an unedited search result in an appendix.

Outputs: This section should contain a copy of the documents identified.

Analysis: This is the section where you look at the publications and identify any recurring issues, any areas of good practice, and any things to avoid. Group your analysis under the following three disciplines irrespective of the source of the information:

- evidence for substantial equivalence;
- evidence that your device will perform as intended;
- evidence that your device is safe to use.

9.8 Format for Formal Clinical Evaluation Report

The format of the overall report is left to the company. However to make it useful for both FDA and CE uses it should be written in such a way that the data can be extracted easily for any purpose. To this end a format is not suggested but it is suggested you follow the guidance in MEDDEV 2.7.1 rev 3 or that laid down in the *Format for Traditional and Abbreviated 510(k)s* (FDA, 2005). Both contain similar information but it is impossible to complete the 510(k) submission without conducting the evaluation described earlier; equally the information used for the 510(k) submission is identical to that one would produce for the MEDDEV report. Hence the following structure may be useful as a starting point.

Section	Content
	Title Page
1	Executive Summary
2	Indications for Use[5]
3	Declarations of Conformity
4	Device Description
5	Classification
6	Proposed Labeling and IFU
7	Literature Review
8	Substantial Equivalence
9	Sterilization and Shelf Life
10	Biocompatibility
11	Software
12	Electromagnetic Compatibility and Electrical Safety
13	Performance Testing – in vitro
14	Performance Testing – animal
15	Performance Testing – in vivo
16	Other

Note: sections 8–15 cover the whole of the evaluation process and will refer back to the literature review and the classification. However some may not be applicable to your device; do not simply discard them. Keep them in the report and justify why there is no content. So, for example, there may be no clinical trial data because the literature review and substantial equivalence have demonstrated that the device is in common use and as such a clinical study is not required to demonstrate safety. Making this statement is just as powerful as having the data itself.

Remember, if you are presenting data in sections 8–15 then the literature review should be referred to in such a way that both support one another, giving your justifications credence. For example, if your screw breaks at 6 Nm (section 13) then the data should be compared with the literature review, which hopefully demonstrates that your device matches or exceeds the norm.

---

[5] Include any contraindications or guidance.

9.9 Summary

In this chapter we examined the clinical evaluation phase of the design process. We saw that it is as important to design this phase as it is to design the product itself. While we recognize that we may not be able to perform all the tasks ourselves we saw that it was important to understand what needs to be done.

References

BSI (2011). *BS EN ISO 14155:2011 Clinical investigation of medical devices for human subjects – good clinical practice.*

BSI (2009). *BS EN ISO 14971:2009 Medical devices. Application of risk management to medical devices.*

Collins (2011). *Collins english dictionary* (11th ed.). England: Collins Sons & Co Ltd.

EC (2009). *MEDDEV 2.7.1 rev 3 clinical evaluation: Guide for manufacturers and notified bodies.* European Commission.

FDA (1997). *Design control guidance for medical device manufacturers.* FDA.

FDA (2005). *Guidance for industry and FDA staff: Format for traditional and abbreviated 510(k)s.* FDA.

MHRA (2011). *Guidance notes for manufacturers on statistical considerations for clinical investigations of medical devices.* MHRA.

## CHAPTER 10

Manufacturing Supply Chain

10.1 Introduction

There is no doubt that the last thing most designers concern themselves with is how a device is going to be made. Right at the very start of this book, we saw that the best way to develop a device is to have the potential manufacturers in at the start. However, one still has to find them. This chapter concerns itself with the rigors of making, maintaining, and regulating your supply chain.

The word "manufacturer" leads to infernal confusion. Under EC rules, as the device specifier and holder of the CE mark you are the manufacturer, as you are in the USA (as the specification developer). However, this does not stop you from subcontracting aspects of the manufacturing process to someone else – but you must do this properly and within a framework.

10.2 Identifying Potential Suppliers

Once again, your quality manual should have a procedure for purchasing. The first part of this purchasing procedure always contains the identification of suppliers. You will not be able to get away with purchasing items at random; under the medical devices regulatory framework you must name suppliers for all critical components/services. In the case of the FDA it is a part of the 510(k) submission, and changing a supplier may make your 510(k) invalid; that is somewhere you do not want to go.

There are no hard and fast regulations, but I can give you some rules to follow. There is no regulation that states who you must go to, but common sense should suggest that you only use a supplier who already provides components and services in your classification. Hence one would not normally use a company to make a Class III (EC) device if they have only ever made items up to Class I.

> *Rule of manufacturing 1: Always use a supplier who has supplied items to others in your classification that are similar in nature to your device.*

If you work with your supply chain you will find that obeying Rule 1 is not difficult. All you need do is ask; most suppliers are more than happy to show off their portfolio of customers and products.

Medical Device Design.
DOI: http://dx.doi.org/10.1016/B978-0-12-391942-7.00010-6

Identification of certification level is of paramount importance. If you are dealing with an ISO 9001 or ISO 13485 registered company then you know that their document trail is going to meet your needs to delegate the responsibility of quality auditing to their notified body.

> *Rule of manufacturing 2: Always use an ISO 9001 supplier as a minimum requirement.*

It cannot be emphasized enough how much Rule 2 will help you. Without this minimum certification requirement you will have to perform full audits of the company in question *yourself.* This in turn means you must be a qualified external auditor. Do you really want this added burden? You may if you are making a very high risk device, but for most companies *audit by certification* is more than adequate.

> *Rule of manufacturing 3: For Class II devices and above always look for ISO 13485 certification or equivalent.*

> *Rule of manufacturing 4: For implants look for a history of implant manufacture.*

Rules 1–4 make perfect sense. They basically state that you should use someone who knows what they are doing. Do not be tempted to cut costs by dropping your own standards. The same rules can be applied to services too; for example,

> *Rule of manufacturing 5: For delivery of sterilization services use a recognized sterile service provider.*

You may think this is obvious, but some sterile packagers only provide a packaging service. Some pack and sterilize. Some will also design your pack, test it, then pack and sterilize. You *must* ensure they have the certificates and track record for the service you want.

10.2.1 Samples

Before you put any supplier on an approved suppliers list you should always obtain samples. This does not mean simply looking in a salesman's briefcase and looking at sales specimens; it means getting them to make something from your device and then holding them to this quality when it goes into batch production.

You may be asking why. The first point is that you want to ensure they can make the item to the quality you want. You will be amazed how the same drawing of a component can result in a variety of finishes, even if it is a complete drawing. The second reason is that any company can produce one excellent item, but can they produce 20 to 1000 excellent items? You can then use this sample as your check for the production runs.

Another good reason for asking for samples is that in your quotation process you will find that suppliers have "sweet spots." Some items they can produce with relative ease, because that is what they specialize in. These will be relatively cheap. Other items out of the "sweet spot" will cost more. And that is why Rule 1 is so important.

10.2.2 Initial Audit

Consider having an initial audit just to ease your mind. This need not be a rigorous procedure but you should supply the auditors with an audit plan (a list of things you wish to examine) and then write a brief report of this audit. You may be surprised by what you find; companies who on paper seem to be excellent can be a letdown when you actually get to the actual premises. Do not forget to examine the auditors' own audit trail. Examples of things you should be looking for are:

- General cleanliness
- Paperwork following products
- Raw material quarantine procedures
- Use of animal products
- Tooling (potential cross-contamination)

As we have touched on earlier, the second to last point is of paramount importance. You *must* get a statement from the company that no animal products are used in their processes. If this is not forthcoming then you really should question their suitability as a supplier unless they are absolutely necessary.

10.2.3 Contractual Arrangements

Every regulatory body will expect you to have contracts detailing your subcontractors' responsibilities. It is a truism that without these you will not survive an audit. The contracts need not be onerous but they must stipulate certain things:

- They will only make to your specification.
- They will not substitute materials without your written approval.
- They will not modify any of your part drawings.
- They will retain relevant documents for the prescribed period, or supply them to you for you to keep.
- They will supply a statement of conformity with each batch.
- They will not use animal by-products in any process, without your knowledge and consent.
- They will notify you immediately of any nonconformance or hazard that they become aware of that may impact on your devices.

In addition, of course, you will have your own performance-based criteria.

Table 10.1: Example Layout for Approved Suppliers List

Company	Contact	Certification	Audit by	Report	Part Numbers
Fred Smith Medical	James Machin 0485 755664 jm@FSM.com	ISO 13485 (expires Nov 2013)	Certificate Audit 14 Jan 2012	FSM1	X-101-1 100-0
JMB Sterile Packaging Inc.	John Brown jbrown@JMB .com	ISO 9001 (Nov 2015) ISO 13485 (Nov 2015)	Certificate	JMB1	All sterile packs. Sterilization services. Accelerated life tests for std packs

10.2.4 Approved Supplier Register

Once you have completed your audit and you are satisfied, the 'potential supplier' becomes an 'approved supplier'. This you need to record in an Approved Supplier Register. The report should record the outcomes of the investigations described above; it should also contain the relevant quality certificates (which must be kept up to date). A typical register entry for a company may be as shown in Table 10.1.

Obviously your register would contain all the relevant certificates and the suppliers list would be an aide-memoire for both annual supplier audits (if required) and certificate updates. You should include an audit of the register as a part of your annual internal audit process.

Note that this register has two further functions. The first is that it stops creative procurement staff from purchasing items from the cheapest source only – this can play havoc with product quality. The second is that it is an information file for your design process – this file tells you "who is good at doing what" and hence who is the best person to bring in at the first stages of a design process. In small companies this is very easy, but when the company gets moderately large this type of information becomes invaluable. It is, after all, a simple *contacts management system*.

10.2.5 Suggested Procedure

You must have a purchasing procedure to meet ISO 13485. We have come across these earlier in the book, however Figure 10.1 illustrates Section 10.2 as a procedure.

10.3 Packaging

Essentially packaging comes within two main criteria. There will always be some form of internal pack that protects the device and (if necessary) its sterility. The second is the outer case that is necessary for transportation and storage. Your packaging selection protocol must encompass both aspects.

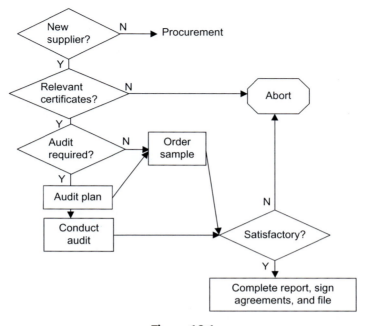

Figure 10.1
Example new supplier procedure.

10.3.1 Sterile Packaging

Sterile packs tend to come in two forms: flexible wrapped and rigid blister. The flexible wrapped kind is the sort of packaging that you would see with any sterile wound dressing you would purchase from a chemist (drug store). This type of pack is reserved for relatively light objects such as adhesive dressings, small bone screws, and giving sets. The blister pack is for heavier objects: ones whose shear bulk would damage the weaker wrappings of the former. For general use the sterile packaging would come single wrapped, that is, there is only one seal between the device and the outside world (Figure 10.2). If, however, the device is going into a sterile environment then it would be double wrapped (Figure 10.3). The simple reason is that the inner pack will remain sterile and can be passed to a sterile operative. If it were single packed the packaging itself would be nonsterile and hence cannot be passed on to anyone in the sterile field. For this reason all devices bound for the operating theater (OR) are, almost exclusively, double wrapped (Table 10.2).

There is a range of materials from which the pouches can be made, including:

- 63 gsm PeelPlus
- 1073B Tyvek
- 12/38 PET/PE
- 60 gsm paper
- 12 mu PET/9 mu foil/50 mu Peel PE

Figure 10.2
Some typical single wrapped sterile pouches. *(Courtesy Riverside Medical Packaging Ltd)*

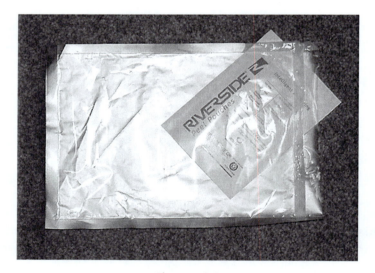

Figure 10.3
A typical double wrapped sterile pouch for use in a sterile field. *(Courtesy Riverside Medical Packaging Ltd)*

But seek advice from a packaging specialist.

The packaging will need protection from the real world, so the design of the outer carrier is of vital importance. Your package may have to travel thousands of miles before it is used and the sterile inner must get to its destination unharmed. For this reason the design of sterile packs

Table 10.2: Example Sterile Packaging Methods

Device	
Light/flexible Flexible wrapping	Heavy/complex shape Blister pack
Use	
General/domestic Single or double wrapped	Sterile field/OR Double wrapped

is best left to those with the expertise and know-how. Most medical device journals have a directory of providers of this service, and of course the web is indispensable. However, follow the rules of approval described earlier.

The packaging and sterilization will have to pass formal approval criteria. There are specific standards depending on your sterilization method. For example, the approval for the evaluation of devices sterilized by irradiation is controlled by ISO 11137; ISO 11135 control approval for ethylene oxide; and ISO 17665 controls steam sterilization. Your packaging must make the fact that the device is sterile patently clear (we shall see this later in labeling); it must also illustrate the sterilization method. But in all cases the packaging will have to pass stringent tests, accelerated life tests, and real-time evaluations before it can be classed as a sterile packaged device. This evaluation is governed by ISO 11607.

Note: your decision to be sterile or nonsterile packed should not be taken on a whim. The cost of producing a document proving your packaging and sterility regime is acceptable is high. It is not just the cost of the test; you could be scrapping up to 80 devices just to conduct the test and the cost of each item can mount up to a scary total. There is also the cost of maintenance; you will have to undertake regular evaluations proving your packaging is still acceptable (normally every three months) and again you may be scrapping many devices in the process. Hence your decision to go sterile must be based on good market intelligence and not just because it seems a nice thing to do. You must have better things to spend your money on.

10.3.2 Nonsterile Packaging

Once again it is highly likely that your device will have an inner and an outer pack. Your inner pack may be something quite special and could easily be your standard inner pouch, as described above, but one that has not been sterilized (Figure 10.4). Take care that your nonsterile items do not look similar to sterile items – this is just too disastrous to contemplate. The benefit of using standard pouches is that all standard details can be preprinted on the pouch itself. If your device is reusable and is to be steam sterilized then you may wish to consider the role of a sterilization case.

A sterilization case is a specifically designed tray, normally made out of stainless steel or anodized aluminum, which retains your device while it is being washed and then steam

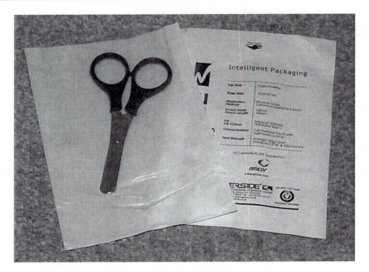

Figure 10.4

An example of a single wrapped nonsterile item. *(Courtesy Riverside Medical Packaging Ltd)*

sterilized. The whole tray, therefore, must contain your device and fit into standard washing machines and autoclaves. There are a number of specialist tray manufacturers who can design something special. Equally you can purchase one "off the peg (Figure 10.5)." However, as stated previously, you will have to prove that your device can be washed and sterilized in this tray…you cannot assume that is the case.

The tray can be fitted with special stands to retain your device securely (Figure 10.6); this also helps theater staff to see if an item is missing. It is quite common to print a manifest of the components onto the tray in their respective locations. This all makes life for the scrub nurses, theater staff, and sterilization staff easier. One of the other benefits of the tray is that it also acts as your protective inner pack for transportation.

Your device will probably need some transportation protection. If you use bubble wrap, etc. make sure it comes from a source that is medical compliant and that you have not inadvertently added a risk of animal by-product contamination.

10.3.3 Packaging Testing

The outer box is now for transportation and storage only. It must be designed to withstand the rigors of transportation. This must also be confirmed. There are several standards for packaging acceptance criteria that you can follow. Remember your main aim is to get "your baby" to the end-user in the state in which it left you, i.e., pristine. Do not allow some delivery courier to ruin your day by rough handling. I promise you that unless you deliver something yourself you cannot guarantee the handling that you would administer. For your

Figure 10.5
Typical off the shelf sterilization tray: (a) simple sterilization case; (b) common device tray.
(*Courtesy Intelligent Orthopaedics Ltd*)

own sake (if only to minimize claims against damaged goods) perform simple acceptance trials such as drop tests and vibration tests. If your device is susceptible to water you should also conduct saturation tests. Table 10.3 gives some example standards.

You may be able to conduct these tests in-house. However some are very specialized and you will have to gain access to recognized testing houses; once again most universities with an engineering department would be able to offer these services.

10.3.4 Storage Considerations

Do not forget that these boxes need to be stored, if not on a shelf in your warehouse then on a shelf in a hospital. When you come to design or select your packaging ensure that it

*   is easily stored on standard shelving;
*   can be handled by one person (<25 kg);

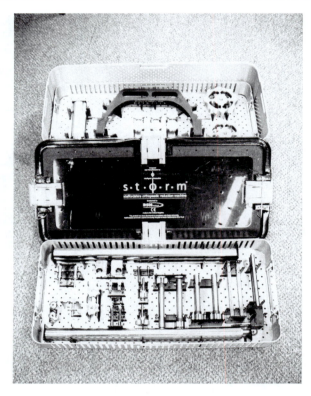

Figure 10.6
Typical bespoke sterilization tray including retainers. *(Courtesy Intelligent Orthopaedics Ltd)*

Table 10.3: Example Packaging Standards

Country	Standard	Description
USA	ASTM D5276	Standard method for drop test of loaded containers by free fall
International	ISO 8318	Packaging filled, sinusoidal vibration tests using a variable frequency
International	ISO 2875	Packaging complete – water spray test
International	ISO 11607	Packaging for terminally sterilized medical devices
USA	ASTM D3592	Standard practice for commercial packaging (note for delivery to U.S. Department of Defense)

- is not one-side heavy or top heavy;
- will not degrade over the period of storage; and
- makes your device easily identifiable in a plethora of similar boxes.

If your device is heavy (>25 kg or >50 lb) you will need to consider lifting/handling arrangements:

- Can it be easily handled by two people?
- Will it need lifting with specialist equipment?
- Will it need to be delivered on a pallet?
- Can the package incorporate wheels for ease of movement?

Your device may move from place to place (such as equipment on a loan or trial basis). In this case your packaging must cope with transportation, packing, unpacking, and repacking. Not only does this take a toll on the packaging but it also tests your package design…after all, do you want to receive a ballooning box that is bulging with its contents just because you forgot to supply packaging instructions to help someone repack it properly?

The last thing to consider is where your device is going to be stored. Will it be

- dry?
- damp?
- hot?
- cold?
- dusty?

The location of an item, while waiting to be dispatched, is often forgotten!

10.4 Procurement

At some stage you are going to have to order your devices. You will now face the sticky question of how many to order. You must work closely with your marketing team to produce a procurement strategy. There are two main pitfalls to avoid:

- having stock on a shelf in a warehouse
- having no stock at all

The former simply ties up liquid assets in stock. While it is nice to see wall-to-wall boxes of stock and say to yourself "look at that, one million dollars sits there," it is better to have sold the one million dollars' worth of stock! Having no stock at all is in fact worse as you will not be able to fulfill orders and this may lead to the embarrassing situation of a cancelled procedure. This you must avoid at all costs.

Why does this affect your supply chain? If you work well with your sales department you can build a good delivery model and hence be able to work with your supply chain in a number of ways. If possible, you may be able to run a "just in time" system with them. But what you must avoid is securing a supplier who can work with batch sizes of 20 per month, only to start ordering 50 per month. This way madness lies. As batch sizes increase,

Table 10.4: Some Common Terms Used by Supply Chain

Name	Description	Batch Sizes
Just in time (JIT)	Goods are delivered exactly when needed and little/no stock held on shelves.	Large, good for regular consumables
KANBAN	Historically a card-based system that "pulls" stock as it becomes depleted.	Any
Batch size	The number of items in any one order.	Any
Materials resource planning (MRP)	A software-based system to control inventory and production planning.	Medium to large batch sizes
One-off production	Literally only one (or maybe two) items are made. Each batch is different to the last.	Very small
Bill of materials (BOM)	A complete list of components, assemblies, and subassemblies for your device.	All
Lead time	The time it takes for the component to arrive after it has been ordered.	All
Constant work-in-progress (ConWiP)	The flow of product is continual with equal size batches arriving at equal intervals.	All, but sales must be highly predictable and repetitive. This is for a product that is well established and has significant market share.

different manufacturing methods become more effective. For example for small batches you may obtain a polymer-based component by machining it directly from a block: however, as batch sizes increase this method of production may become costly and you may wish to change to injection molding. All of this needs planning, and you will be unable to solve this issue overnight. So the message here is plan well in advance and secure your short-term, medium-term, and long-term suppliers well in advance. If you do not do this your purchasing department will grow to hate you.

10.4.1 Supply Chain Glossary

Table 10.4 illustrates some common manufacturing terms you may meet when discussing procurement with your supply chain.

10.4.2 Costing

Costing is an art. Unless you have spent your life costing the manufacture of components, this is a lost cause. However, most CAD packages come with an estimation algorithm built in to the system. Hence it is now relatively easy to estimate what the manufacturing cost of a component is likely to be.

Remember, the CAD system will never be able to replace the art of negotiation.

10.5 Summary

In this chapter we considered the supply chain. We saw how we had to have a robust selection process culminating in an approved supplier register for all of our key suppliers. We examined one of the major suppliers, sterile packagers, and we met various forms of packaging methods. Finally we examined the role of modern manufacturing technologies in a modern medical devices framework.

Further Reading

It is beyond the scope of this textbook to make you into a manufacturing engineer. However it is a truism that your gross margin will be dependent on the reduction of your manufacturing costs, hence it will do you little harm to read further. Some suggested texts from which you can start are:

Bicheno, J., & Catherwood, P. (2005). *Six Sigma and the quality toolbox*. PICSIE books.

Liker, J. (2004). *The Toyota way, 14 management principles from the world's greatest manufacturer*. McGraw-Hill.

Vollman, B., & Whybark, (2004) *Manufacturing planning and control systems for supply chain management: The definitive guide for professionals*. McGraw-Hill.

## CHAPTER 11

Labeling and Instructions for Use

11.1 Introduction

You should be quite used to me telling you "this is a part of your design"; and so it is for these two important items. You need to remember that your device is going to be used by someone who has never even met you let alone seen *you* use your device. Hence your labeling and *instructions for use* (IFU) must make the whole process from delivery to final use as easy and as stress free as possible. The only way to achieve this utopian outcome is to think about the design of your labels and IFUs from the start. We will, however, see that some of our work has been done for us by standards and guidelines. This chapter will be split into three main sections: labels, IFUs, and surgical techniques. But we will see that they are all interlinked and that there are some classic traps that can be avoided with a little bit of forethought.

11.1.1 The Rules

All regulatory bodies have a section of their rules and regulations dedicated to correct labeling of your device. It is impossible to demonstrate every possible permutation for every possible country of sale – but it is possible to lay down some basic design rules.

The FDA and EC guidelines for labeling (and marking) are quite clear and well documented. You should download these documents and adhere to them.

- Label: a mark or printed label that is attached to (or printed on) an external package. It is not permanently affixed to a device.
- Marking: indelible/indestructible marks on a device used for recognition of said device a long time after its initial use.

Table 11.1 lists some of the plethora of documents that will help you to design your labeling and marking strategy. Unfortunately, the level of information you need to supply varies with the classification of the device. Table 11.2 attempts to illustrate the level of complexity that this section of your design process hides. Your PDS should have addressed labeling and marking requirements. The same methodologies of idea generation and selection we met earlier in this book should be used, as with the device itself. This will help make sure that your labeling is foolproof.

Medical Device Design.
DOI: http://dx.doi.org/10.1016/B978-0-12-391942-7.00011-8

Table 11.1: Standards and Guidelines Associated with Labels and Marking of Medical Devices

Region	Title	Comment
USA	21 CFR Part 801 General device labeling	Freely available on FDA website
USA	21 CFR Part 812 Investigational device exemptions	Freely available on FDA website
USA	Guidance on medical device patient labeling; final guidance for industry and FDA reviewers	Freely available on FDA website
USA	Use of symbols on labels and in labeling of in vitro diagnostic devices intended for professional use	Freely available on FDA website
USA	Alternative to certain prescription device labeling requirements	Freely available on FDA website
WO	ISO 15223: Medical devices. Symbols to be used with medical device labels, labeling and information supplied	Available online at a cost
EC	Medical Devices Directive	Freely available online.
EC	The CE Mark: MHRA Bulletin No. 2	Freely available on MHRA website
EC / UK	BS 3531-6: plants for osteosynthesis. Skeletal pins and wires. Specification for general requirements.	Available online at a cost

Table 11.2: Labeling and IFU Needs for Specific Classes of Device

Requirement Device	Labels											IFU					Device				
	CE Mark	Rx Symbol or Statement	Sterility Symbol	Single Use Symbol	Nonsterile Symbol	Lot/Batch Number	Packing Date	Use by Date	Manufacturer Details	Device Identification	Patient Labels	Certificate of Cleaning and Sterilization	Cleaning and Sterilization Instructions	IFU Sheet	Declaration of Conformity	Surgical Technique	CE Mark	Device Identification	Lot/Batch Numbers	Single Use Symbol	Manufacturer Identification
Class I nonsterile item	✔			✔	✔	✔	✔		✔	✔		✶	✔	✔	✶	✔□	✔	✔	✔	✔	
Class I nonsterile – reusable device	✔	✔		✔	✔	✔	✔		✔	✔		✶	✔	✔	✶	✔□	✔	✔▲	✔▲		
Class II nonsterile item	✔*	✔		✔	✔	✔	✔		✔	✔		✶	✔	✔	✶	✔□	✔	✔	✔	✔	✔
Class II single use sterile item	✔*	✔	✔	✔		✔	✔	✔	✔	✔	✔			✔	✶	✔□	✔	✔	✔	✔	✔

▲ marked on each separate item; ✶ provided on request; * includes Notified Body registration number; □ if risk analysis requires one.

11.2 Standard Symbols and Texts

For all markings and standard symbols refer to the most current guidelines and standard applicable for the country of sale. The following have been derived from ISO 15223 (ISO, 2007) and from wide experience. Always check with the relevant regulatory authority first. However, if you work closely with your end-users they will have examples of good practice from other companies in the medical devices field: *on the shoulders of giants!*

11.2.1 CE Mark

For all devices for sale in the EC, there must be a CE mark (as per EC guidelines) on the device (Figure 11.1). This need not be on each component, but on each individual device. For all devices of Class II and above the CE mark must also contain the Notified Body's number.

There is no such FDA equivalent. The FDA does not license a device; they only give a company the *"clearance to market"* a device.

11.2.2 Nonsterile Device

Figure 11.2 illustrates the standard nonsterile symbols for use in the EC and the USA.

11.2.3 Single Use Item

Figure 11.3 illustrates the standard accepted symbol for single use only.

Figure 11.1
CE marks (use the format given by the EC): (a) Class I device; (b) Class II and above.

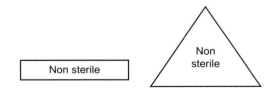

Figure 11.2
Nonsterile symbols: (a) EC – pre ISO 15223; (b) USA and EC ISO 15223.

Figure 11.3
Single use symbol.

11.2.4 Sterility

This has a number of symbols; they all depend on the method of sterilization (Figure 11.4). The symbols look the same; it is the last letter that differs.

11.2.5 Use by Date

Please note that dates have a very specific format. They always go year – month – day. Hence 1st February 2012 would be 2012-02-01 (Figure 11.5). Stick to this format and never ever miss the zeroes!

11.2.6 Lot Number/Batch Number

In this case XXXXX represents the unique lot number (or batch number) for this device (Figure 11.6). It is the single piece of information that ensures *traceability*.

11.2.7 Catalog Number/Part Number

Here, XXXXX is the part number for your device (Figure 11.7). It helps with traceability, but also for reorders!

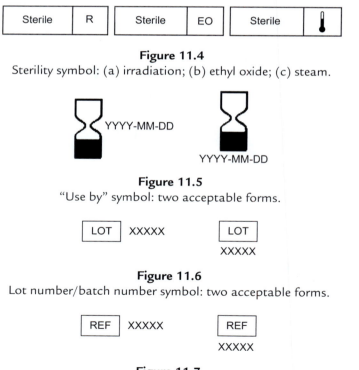

Figure 11.4
Sterility symbol: (a) irradiation; (b) ethyl oxide; (c) steam.

Figure 11.5
"Use by" symbol: two acceptable forms.

Figure 11.6
Lot number/batch number symbol: two acceptable forms.

Figure 11.7
Part number/catalog number symbol: two acceptable forms.

Figure 11.8
Consult instructions for use symbol.

Figure 11.9
Caution symbol.

doc: foolscap/2003 rev 1.2

Figure 11.10
Caution symbol pointing the user to read "foolscap/2003 rev 1.2" before use.

11.2.8 Consult Instructions for Use

Both of these are very important symbols. Figure 11.8 appears on every label of every device. It says "the end-user should read the IFU before using." Figure 11.9 is a caution; this is often used where there are surgical techniques in which there are particular precautions. This is especially useful if you have a surgical guide that is available; point the user to this by putting the document reference under the symbol (see Figure 11.10).

11.2.9 Prescription Only

In the USA "prescription only" means something different to when it's used in many European countries. In the UK, for example, "prescription only" means the doctor/surgeon/ clinician needs to write an actual prescription and the device will need to come from the pharmacy. In the USA "prescription only" means it can only be used by or under the instruction of a clinician: it is an essential requirement for your label if your device cannot be used by the general public. Hence, be careful when using the recognized Rx symbol – it will cause problems in the EC. If you are going to sell in the USA then there are accepted words that mean the same thing, but do not affect your device in the EC:

Caution: U.S. federal law restricts this device to sale by or on the order of a physician.

You must have this wording on your device in the USA or you will not obtain your 510(k). This statement avoids using the U.S. symbol Rx[1] , which (as I said earlier) will cause grief in the EC.

11.2.10 Manufacturer Details

This symbol is very important and the information itself should be easily found on all of your documents (Figure 11.11). If there is any issue with your device the end-user must be able to contact you, and contact you quickly. This address and the web address must enable the end-user to get in touch with you straight away. If it does not, then you need to rethink your contact details.

11.2.11 Date of Packing/Manufacture Date

Once again the date is in year – month – day format (Figure 11.12). This piece of information is essential for efficient traceability!

11.2.12 EC Representative

If you are trading in the EC but are not an EC-based company then you need to state who your representative is. This is normally on the packaging or within the IFU itself (Figure 11.13). Again, this is needed as a point of contact in the case of emergencies.

Medical Devices Inc
402 West Virginia Rd
Stoke on Trent
UK ST4 5ST

www.mdinc.com

Figure 11.11
Manufacturer's contact details, including the now commonplace web address.

YYYY-MM-DD   YYYY-MM-DD

Figure 11.12
Symbol giving date of manufacture (normally date of packing): two acceptable forms.

Joe Smith Medical
Friend on Dee
UK DF10 1EZ

Figure 11.13
EC representative symbol.

---

[1] Rx is purported to originate from the ancient Egyptian symbol for the *Eye of Horus*. It was used to call, ask, or pray for support against ailments or to seek healing. Nothing new under the sun!

11.3 Labeling

All we need do now is put all of the symbols into a meaningful layout. I am afraid this is down to your own thoughts and guidelines. However you can make life easy for yourself by incorporating things that do not change on permanent artwork, and those that do change using self-adhesive labels. At the end of the day the choice is yours. Certainly for small numbers the expense of preprinting boxes and packages with standard livery is just not cost effective and you will, almost certainly, be supplying Avery Inc. with good business. But as your numbers increase, moving to standard livery is extremely useful.

When you start to print your labels you will quickly find that the cheapest labels in the high street will not do. The adhesive used on cheap labels does not perform well and you will soon have complaints from your end-users that labels have peeled off and, as a consequence, they are unable to use your product. Self-adhesive labels are too important to scrimp on. The world of labels is populated with a plethora of sizes. A good piece of advice is to pick one label size (or at most two) that fits all of your products; this reduces inventory but also enables you to design a standard layout. Pick an efficient and effective manner of printing. You should not scrimp on printers either; the text, symbols, and any bar codes must be legible so a good quality laser jet or one of the continuous thermal printers (e.g., Dymo TurboJet) is essential. Once you have got this information labels are simple.

11.3.1 Outer Packaging Labels

The labels need not be beautiful, colorful, or able to win design awards. They only need meet the labeling requirements for the given country, or state. Figure 11.14 illustrates a typical simple label. It can be printed off with ease but only has three items of information that change: those ringed. The more eagle-eyed of you will have noticed three deliberate mistakes. The manufacturer symbol is missing (11.2.9), and the address is incomplete. The REF symbol associated with the part number (11.2.6) is missing, and this label uses the "old" nonsterile symbol.

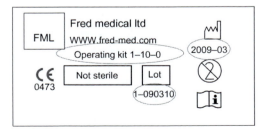

Figure 11.14
Example label for EC: contains intentional errors.

Remember: if your device has separate inner packaging the label must be replicated on any inner packaging. This is especially important if your outer packaging is a multipack. If this is not undertaken traceability will be lost.

11.3.2 Patient Labels

It is common practice, especially if your device is for use in theater, to supply patient labels. These are small self-adhesive labels (normally 5) that can be peeled off and attached to the patient's notes (Figure 11.15). This makes life easier for the clinical and OR staff. The label usually contains only the bare minimum information such as part number, batch/lot number, and any other information deemed important to identify the device. Once again this is useful to enable efficient communication should anything remiss happen. These labels are quite common for sterile single use items.

11.3.3 Bar Codes

Many hospitals demand bar codes for inventory control. Normally they require three items of information: part number (catalog number), batch/lot number, and packing date. For sterile/ perishable items this may also include a "use by" date. You should find which bar code system the hospitals want; you should also have a printer and software that can print bar codes (many thermal printers such as the Dymo TurboJet series come with this software as standard). It is quite common to use the UCC/EAN128 bar code profile. It is also important to use the right

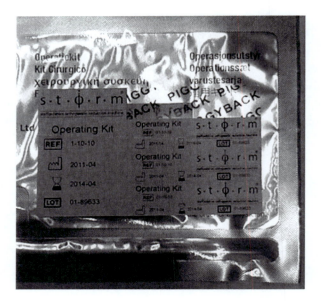

Figure 11.15
Example patient labels. *(Courtesy Intelligent Orthopaedics Ltd).*

configuration codes. Figure 11.16 illustrates a typical bar code label for a Class II nonsterile product. It has three main bar codes. The first is the manufacturing (packing) date – the number in brackets is the universally recognized code (called a UCC identifier) for this item of information. The second bar code is the batch/lot number (10), and the third is the part number (code 241). As per normal, your PDS will have identified which bar code style to use.

I do not know why, but it seems to be standard to affix the bar code on the back of the box in the exact middle. However, I find this to be a silly idea; if your device is fragile you are introducing a potential breakage just for identification purposes. Put the bar code label where it is easy to get at (how many times have you been at a checkout watching the store staff struggle to find the bar code for your purchase), and where its use will not cause damage.

11.3.4 Security Labels

Your packages should be made tamperproof. Hence it is a good idea to produce a standard label that states:

Do not use if packaging damaged.

Use this as your final seal for the box/pack.

If your device is supplied sterile this label/statement (or wording like it) is mandatory.

11.3.5 Crossover of Symbols between USA and EC

Although standard symbols are commonplace in the EC, the same cannot be said for the USA. EC companies selling in the USA have two choices: produce USA packaging and labels, or (for small numbers) get a "meaning of symbols" label agreed to by the FDA examiner. For USA companies there is no alternative but to use the agreed symbols in the

Figure 11.16
Sample bar code label (UCC/EAN 128).

EC. As with all such queries, obtain guidance from the regulatory bodies before undertaking the expense of printing!

The sooner this aspect of medical devices regulations gets harmonized the better. Luckily, there is consensus starting to form and as soon as ISO 15223-2 (ISO, 2010) becomes live, all labeling will become a lot easier.

11.3.6 Translation

The benefit of symbols is that they need no translation. However if you need to have a product description, such as *Giving Set*, then this will need to be translated into the languages of the countries into which you will sell. Even a modest global company can expect to have 13 translations (English, French, German, Spanish, Japanese, etc.). Hence try to avoid unnecessary translations as these will not only take up valuable label space but will also cost you in translation fees. We will discuss translation protocols later.

11.3.7 Position of Labels

It is commonplace to have labels on the top and bottom of the box. However, think of the nurse or technician who has to find the box on a stack of shelves. For them the better place for a label is on the end of the box, which can be seen without disturbing the shelves. Once again, your customer requirements (in the PDS) should have highlighted this issue.

11.4 Marking

Your device *will* need permanent marking. For safety's sake (and this is compulsory in many countries) each component (that can be removed from said device) needs to be individually marked. Why is this a necessity? Quite simply your device's original packaging will, probably, be destroyed at the point of opening. Hence any record of its origin will be hard to find. All end-users must be able to determine your device's part number and lot number at any time of its life cycle. Hence permanent marking is important – but why for each individual component? Things can get lost in cleaning, in unpacking. etc....how is anyone supposed to identify your component if it has no identity?

11.4.1 Company Identification Mark

It is very common for a company to have a recognizable trademark. In medical devices this has another currency than simple brand awareness: instant and obvious recognition of the manufacturer. One must always consider the potential for a device to fail. Once your packaging has been removed and disposed of how will someone identify you as the manufacturer? If you are famous in the field then your trademark may suffice, if not you will need your company's registered name.

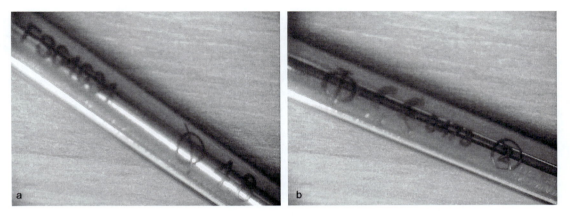

Figure 11.17
Example markings on an orthopedic bone drill: (a) CE mark and single use symbol; (b) lot number and diameter. (*Courtesy Intelligent Orthopaedics Ltd*)

11.4.2 CE Mark

For all devices to be sold in the EC, a CE mark is required. For Class II devices and higher this is the CE mark and the Notified Body's number, as in Section 11.2.1. Note: there is no equivalent symbol for the USA.

11.4.3 Part Number and Lot Number

All items must permanently indicate their part number and their unique lot/batch number. The main device should have an indication of the manufacturer permanently marked too (normally the trademark symbol).

11.4.4 Size

For devices of specific sizes it is very common to mark the size in a recognizable form. For devices that have to be a specific size to work this is essential.

11.5 IFUs and Surgical Techniques

This is often a bone of contention with people outside of medical devices. But the best way to consider the differences between the two is to consider buying some domestic goods. If you were buying a television then it would come with its own "instruction manual" – telling you how to install it and use it – and you would probably make sure it is kept somewhere safe (just in case). If, however, you were buying a pack of wood screws they would probably come with some brief printed instruction sheet stating what they can be used for. In medical devices we must always provide an IFU (instructions for use leaflet); however, depending on the complexity of the system an instruction manual – or surgical technique (you pick the appropriate title) – may be required.

11.5.1 Instructions for Use Leaflet

This is the most common form of IFU for all devices. It is normally brief and will state what the device can be used for, its indicative use, and any contraindications (where it is not to be used). It will have the relevant CE mark (if in the EC) and state whether the device is supplied sterile or nonsterile. It should also give any warnings required (from your risk analysis). The IFU must also state the single point of contact for complaints, reporting of hazards, noncompliance or vigilance. Some countries (Canada, for example) have specific requirements in relation to point of contact so make sure your IFU meets the requirements for the country in which you intend to sell.

If you are able to, use your end-users to supply you with examples of IFUs from their other suppliers. Some medical device suppliers have them as live documents on their websites. You will be able to build your own picture of what is required.

The IFU is a controlled quality document so will need the standard quality tagging (document revision, etc.); it will also need formal approval. Hence it is worth having a standard IFU that you can complete when new products/devices come along. A typical format is shown in Figure 11.18.

Note that in the EC a single use item's IFU must include a statement as to why it is single use. This may sound silly, but its intention is to stop manufacturers from making items single use when they do not need to be – and hence drive up sales, sneakily.

11.5.1.1 Nonsterile Items

Most nonsterile items will need to be cleaned and sterilized before use. Your IFU must describe how this is to be done and which sterilization method is recommended. If your device is to be steam sterilized make sure you state the harmonized steam cycles (ISO 17665-2:2009) within your IFU (Table 11.3). If you do not include these instructions you will find your device will be returned to the clinician from the sterilization center with a rather nasty letter, and you will get a very nasty phone call. It is also without question that anyone inspecting your systems will ask for the IFU and will look for this.

Whatever you do, do not specify a process that a hospital or clinic cannot achieve. This is no use to anyone.

11.6 Surgical Technique

For items that are more complex it may be necessary to produce a more detailed document describing how to use your device correctly. This is called a *surgical technique*. This is equivalent to the operating handbook you would expect to get with a new DVD player. It is common to supply one (or more) with the original first delivery. For more complex devices,

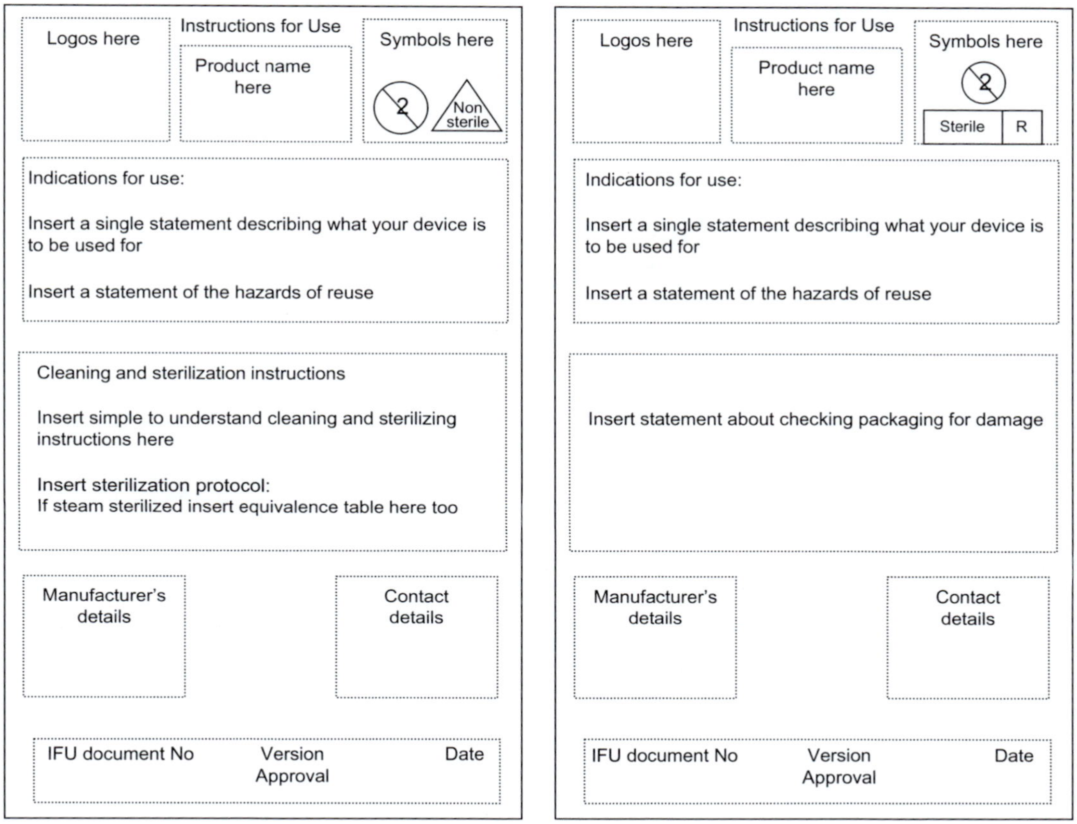

Figure 11.18
Example IFU pro forma: (a) nonsterile single use item; (b) sterile single use item.

Table 11.3: Standard Steam Sterilization Cycles (ISO, 2009)

Temp (°C)	Duration (Minutes)
121	15
126	10
134	3

you may need to supply one per device. There is no rough and ready rule to follow; the choice is yours. However, if in doubt supply one copy per device. Do not forget you must still include the contents of the IFU described in the previous section.

Nowadays, it is commonplace for these documents to be in a PDF format that can be downloaded from your website.

11.6.1 Assembly and Disassembly Instructions

If your device is to be assembled at point of use you will need to supply assembly instructions. If your device is to be assembled sterile then remember that the instructions will need to go into the sterile field and so should be cleanable and easily wiped; paper copies will not do.

It is when you come to develop the assembly instructions that you start to appreciate all of the effort you put into the design of the device earlier on. If you *really* considered the end-users' comments your assembly will be easy to describe….if you did not the assembly process will be horrendous. So a word of warning: do not forget this important aspect in the life of your device.

If you are to base your assembly and disassembly instructions on anything in particular, then IKEA is a good example. They manage to get people to assemble complex furniture with no training at all. What makes their instructions even better is that they are all done with pictures – no translation required!

Do not forget to include a manifest of all of the components in your instruction sheet. All hospitals count in and count out during a procedure, so it helps if they know what they are counting.

11.6.2 Warnings and Contraindications

The surgical technique document is a good place to provide any warnings and contraindications. Why is this so? It is more likely that the practicing clinician will read the technique before using your device; they will (in all practical environments) not look at the IFU sheet.

Contraindication: Any patient profile or situation where the device MUST NOT be used.

Do not provide blanket contraindications – this can limit the scope of your device. Concentrate on your risk analysis and make sure that your device does not compromise any situation where the risk is unacceptable.

A typical contraindication may be:

This device is not to be used in minors or persons who are not skeletally mature.

Warnings: Any patient profile or situation where the device, if inappropriately used may be hazardous.

Once again, your risk analysis will highlight what these contraindications should be.

A typical warning may be:

> *Take care when using the device with minors or patients who are not skeletally mature as excessive traction forces may damage a growth plate.*

You may notice that the two examples above are both sides of the same coin. The contraindication bans anyone from using your device on this patient group, even if they think it is possible. Anyone not following this contraindication is playing career suicide. The warning, on the other hand, says they can use it but if they do and they damage a growth plate (i.e., cause an injury) then it is their fault for not taking care.

Do not forget basic warnings. If your device is sharp, tell them it is sharp! Do not rely on common sense – common sense does not exist.

11.6.3 Production of the Surgical Technique

It is without question that this document needs to be written with close cooperation of the end-user. Even better, a professional end-user should write it. So if your device is to be used by an oncologist then an oncologist should write it; if it is for a staff nurse…and so on. You will find most companies obey this simple rule.

This is a very important document so do not rely on the first draft being correct. You should follow the same design procedures, outlined earlier in this book, for its design and production. Only that path results in the first draft being close to the final outcome. As with the real device, this document needs evaluating before release. Hence have a group of end-users ready to go through the draft and make suggestions.

The final suggestion: a picture paints a thousand words. All technique manuals abound with a plethora of images. In some cases photographs are used, in some drawings…but images are always there to make a description obvious.

11.6.4 Document Control

The surgical technique is a controlled document, so it should contain all of the relevant quality tagging one expects to see with all quality documents.

11.7 Declarations

11.7.1 Declaration of Conformity

Wherever you sell you will need to produce a declaration of conformity. This is a document that sits in your product's technical file and can be produced on request. It is a highly specific document and the regulatory bodies give substantive guidance on what they should contain.

So once again refer to the regulatory bodies for advice. They are controlled documents so they need a version number, etc.

11.7.2 Declaration (or Certificate) of Cleaning and Sterilization

Devices that are supplied nonsterile and which are intended for sterilization on site will need a document declaring that they have passed a cleaning and sterilization trial. Again, this is a controlled document.

11.8 Translation

It is inherent, in most English speaking countries, that the thought of producing a document in anything else but English is downright silly. Surely everyone reads English as standard. No! There is a big wide world out there, and we are not allowed to use that as an argument. There is a regulatory need for you to translate documents into the native language of the country in which you sell. So if you intend to sell into Latin America, Spanish is obvious. If you intend to go further south into South America, Portuguese comes into play. If you intend to sell into the EC then, well, pick a language!

Whatever the language you pick you must have a translation procedure in place, and stick to it. Figure 11.19 illustrates a typical translation procedure.

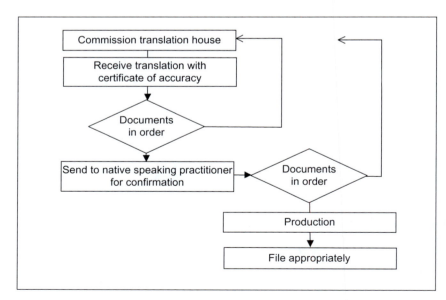

Figure 11.19
A typical translation procedure.

Do not forget, the translation house has become a critical supplier. Hence they must have followed the approved supplier procedure described previously. You cannot pick any old translator. Note: if this is a translation for the Japanese market you must use a Japanese translation house…you cannot use one based in your own country.

Please take extra special care with your labels: these will need translating too. But the text on labels can lead to weird translations. So make sure your professional end-user looks at these as well. The last thing you want is a "nova" issue.[2]

11.9 Summary

In this chapter we looked at the essential documents you need to produce. We looked at labeling, packaging, and additional documents. We saw the importance of the IFU and its sister document the surgical technique. To finish we recognized that translation is an important activity that needs to be conducted with a recognized translation house and the help of your end-users.

References

ISO (2007). *ISO 15223-1 Medical devices – Symbols to be used with medical devices, labeling and information required – General requirements.*

ISO (2009). *ISO 17665-2 Sterilization of healthcare products—Moist heat—Guidance on the application of ISO 17665-1.*

ISO (2010). *ISO 15223-2 Medical devices – Symbols to be used with medical devices, labeling and information to be supplied – Symbol development, selection and validation.*

---

[2] In the UK the Vauxhall motor company produced a new car they called the Nova, which was intended for sale across the EC. They forget to check its translation and hence missed the point that in Spanish it means "doesn't go!" The same problem applied to the Toyota MR2 but with more embarrassing issues for French-speaking peoples.

## CHAPTER 12

Postmarket Surveillance

12.1 Introduction

It is not the aim of this chapter to teach you how to set up your vigilance system; this is left to other texts and documents. It is enough to say that if you do not have one in place then you are either a Class I exempt manufacturer who has never been audited or you are not a medical device manufacturer. In all cases, no matter what class of device, you must have a *postmarket surveillance*, complaints, and vigilance system in place. Those of you who have them working probably think of them as an annual pain when audit time comes round. However, to the designer they are the fountain of all knowledge – really!

In this chapter, therefore, we shall be looking at PMS (postmarket surveillance) differently to how an auditor would – we shall be using it as a quality tool. Remember, many chapters ago, I introduced you to the film *Chitty Chitty Bang Bang* and its song:

> *"From the ashes of disaster grow the roses of success."*

Recall also the wonderful quote of Edison:

> *"Genius is 1% innovation and 99% perspiration."*

There is so much knowledge associated with hindsight that proverbs and sayings abound:

> *"Three failures and a fire a Scotsman's fortune make."* (anon)

> *"We often discover what will do, by finding out what will not do; and probably he who never made a mistake never made a discovery."* (Samuel Smiles)

What is it we learn from these quotes? Firstly we *will* make mistakes, but these mistakes only help to make our designs better. Secondly, people *will* find fault with our designs, but these will lead to improvements and even new products. What is it we learn? We learn that everyone is an expert designer and they will always give you design suggestions – even when you do not want them. Some will do this nicely; others will do this using your complaints procedure. However they do it, you, as the designer, you must capture people's design suggestions. To this end this chapter is all about capturing the information; and this is not as easy as you think!

Medical Device Design.
DOI: http://dx.doi.org/10.1016/B978-0-12-391942-7.00012-X

12.2 PMS and Its Role in Design

As stated earlier, *all* medical device companies *must* have an active postmarket surveillance system in place. Some think this is only to capture complaints; it is not. The PMS is to capture information about your device after it has hit the marketplace, be that good, bad, or indifferent. It is common, therefore, to have a three-pronged attack. The first prong is to have a clinical lead looking at the clinical literature and knowledge base. The second prong is to have the marketing manager collect all information from your sales and marketing staff, and current market literature. The third prong lies with the technical director whose job it is to collect all quality related information and material from the technical knowledge base. The trick is bringing all three prongs into one outcome. Hence all three strands need examining for every one of your main products with a view to coming out with one of the following design outcomes:

- New product
- Design modification
- No change

This can only happen if you discuss all three inputs relative to your device. This need not be every week but it should be more than once a year! A further input is, of course, the emergency input for a design modification – the dreaded Preventative Action Notice.[1]

Figure 12.1 illustrates a typical PMS meeting procedure. The three main areas – technical, market, and clinical – would collect data and bring that data to the meeting. The meeting would then discuss the data to decide which of the three outcomes is the most appropriate. The hard part of this procedure is to ensure that everyone collects the data and does not leave it to the day before!

12.3 Tools

In this section I intend to illustrate some tools that may be of use in your PMS. It is impossible to present all possible tools but these are ones I think are pertinent and easily adopted.

12.3.1 Process Control Chart

This is probably one of the oldest tools around. Primarily it exists to capture slowly changing parameters before they grow into parametric defects. I have already presented to you the need for continual evaluation of your products; this is a simple extension of that process. For example, you have a product and deemed it necessary to randomly inspect samples from every batch. You could simply file that inspection report with the batch documentation and from thence it will only see the light of day if it is ever the subject of an audit. Or you could use it as a part of your continual improvement process by passing the information onwards.

---

[1] A preventative action leading to a design modification would be called if there is a design fault that can be rectified quickly without the need for a recall.

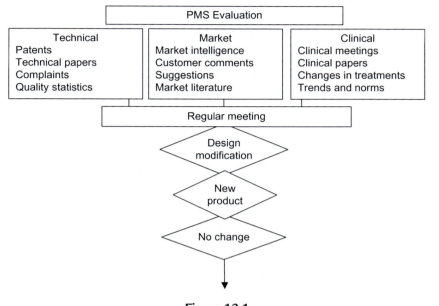

Figure 12.1
A typical PMS outcome meeting.

A process control chart captures this data and plots the information sequentially. Consider the data that normally exists at the foot of every patient's hospital bed: the temperature graph. The nurse will take your temperature every so often and plot this on a chart; if the graph is on a constant rise warning bells ring (not literally!). It is a very graphical way of illustrating errors that vary with time. Your process control chart must have your upper and lower limits of acceptance and then the data is plotted sequentially. Figure 12.2 illustrates a typical chart taken from inspections of the bio-burden of a device before it is packaged.

The main aim of Figure 12.2 is to demonstrate the power of this type of chart, even though it is simple. The horizontal axis is the sample, sequentially. The vertical axis is the measurement taken (in this case it was the bio-burden of a device after manufacture). The vertical line is only there to show that something has changed. Prior to this sample the values of bio-burden had no real pattern; they varied between upper and lower limits randomly. However, after the line it is clear to see that there is something going on. The bio-burden seems to be on a steady increase. We can use this information to investigate before failure occurs. The instigator for this growth could have been a machine change, a personnel change, or simply a lack of attention to detail. In this case it was due to a new machinist who was taking the drawings literally and not finishing the component to the same level as their predecessor and hence creating a place for bio-burden to stick. This led to a drawing change

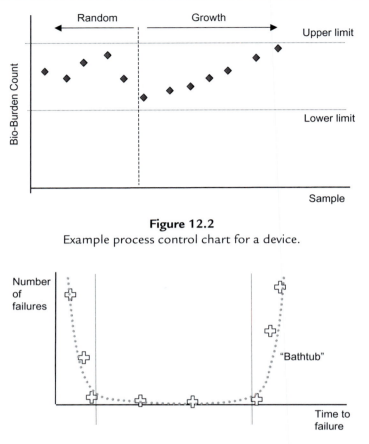

Figure 12.2
Example process control chart for a device.

Figure 12.3
Typical bathtub curve of failures.

which not only led to normal bio-burden results but also a more consistent finish on all components.[2]

12.3.2 Reliability – Bathtub Curve

The Weibull function is over 60 years old – it was established in the 1950s. Ever since, it has been used been to analyze and predict failures. However, there is a more simple process one can use initially and this is the life bathtub plot. For simple components (especially those solely electronic based) the life follows a specific shape – as illustrated in Figure 12.3.

---

[2] You would be amazed at the occurrence of this type of issue. All is fine because "George" has always made things brilliantly, but on his retirement the new guy works from the existing drawings – which did not match George's delivery…they had never needed to. The drawings or component specification must match the expected outcome.

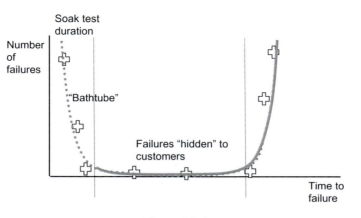

Figure 12.4
Bathtub curve after "soak testing" initiated.

These types of devices tend to fail early on in their life (due to component failure and poor assembly). Those that survive this early period tend to last "forever," but they start to fail as they get old; and this tends to be after a specified period of time. The points tend to make a "bathtub" shape. Televisions, washing machines, etc. tend to illustrate this type of curve – and that is what a 12-month guarantee is for! Companies overcome this by instituting soak testing (running the devices for the specified period to overcome the first part of the curve). What everyone is trying to achieve is illustrated in Figure 12.4.

12.3.3 Weibull Plot

The Weibull (pronounced "vibull") plot is arguably one of the most commonly applied techniques for predicting failure (Carter, 1986). The main reason for this is it works – irrespective of the arguments about its statistical validity. We shall not consider the statistical proofs – that is best left to more worthy texts; instead we shall consider its application. The basis for the analysis is the Weibull plot, and this is illustrated in Figure 12.5. To best describe the plot we shall first discuss the data required.

The Weibull analysis depends on the collection of failure data in the form of "time to failure," "uses to failure," or "cycles to failure." Table 12.1(a) depicts some typical data as collected by your PMS process.

The data in Table 12.1(b) has been ranked in ascending order. The mean rank is determined using Equation (12.1) – much the same as we did for 2^k experiment design:

$$R = i/(N + 1) \tag{12.1}$$

where i is the rank number and N is the number of data points. In the case of equal ranking do not forget to increase the rank at the next change.

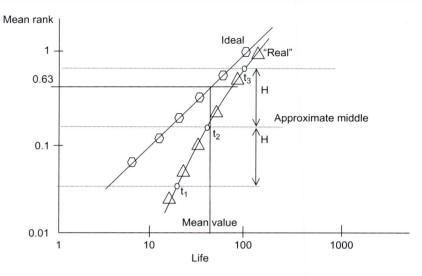

Figure 12.5
A typical Weibull plot.

Table 12.1: Typical PMS Data for Weibull Analysis

(a) Original Data		(b) Ranked Data	
Sequence	Life (No. of Uses)	Life (Ranked)	Mean Rank
1	150	75	1/9 = 0.11
2	100	78	2/9 = 0.22
3	75	95	3/9 = 0.33
4	200	100	4/9 = 0.44
5	125	125	5/9 = 0.55
6	78	150	6/9 = 0.66
7	95	200	7/9 = 0.77
8	215	215	8/9 = 0.88

The next step is to plot this on Weibull paper. You can purchase Weibull paper, or simply generate your own. It is a log-log plot: the vertical axis being the mean rank, and the horizontal axis life.

Figure 12.5 illustrates a typical Weibull plot. The ranked data would be plotted on the graph – note both axes are logarithmic. Ideal data would lie on an exact straight line; however this is often not exhibited by real data. We must straighten the line. To do this we determine a value called t_o, and to do this we use Equation (12.2).

$$t_0 = t_2 - \frac{(t_3 - t_2)(t_2 - t_1)}{(t_3 - t_2) - (t_2 - t_1)} \tag{12.2}$$

The ideal is then produced by plotting life as $(t-t_o)$ instead of life alone. The significance of t_o is that this is the life below which the device can be called intrinsically reliable. The value of life when mean rank is 0.63 is the mean life (but only for the straight-line ideal); if you had to straighten your line the actual mean life is given by Equation (12.3).

$$\text{Actual mean} = t_\text{o} + \text{Mean life(measured@0.63)} \tag{12.3}$$

Clearly this is invaluable information to the designer. If you have designed your device to last for 100 uses then t_o had better be >100! Let us reexamine the data in Table 12.1 and plot these on the Weibull plot (Figure 12.6).

Using Equation (12.2) we obtain t_o:

$$t_0 = 100 - \frac{(180 - 100)(100 - 80)}{(180 - 100) - (100 - 80)}$$
$$= 73 \text{ uses}$$

We now replot the data using the modified life values (t–t_o) (Figure 12.7).

If we now use the 0.63 mean rank and take the corresponding life value this gives us the estimate of average life. In this case it corresponds to about 72 uses. To obtain the estimate of actual average life we use Equation (12.3):

$$\text{Actual average life} = 73 + 72 = 145 \text{ uses}$$

The beauty of this type of analysis is that you can use your service data to constantly update these three values. Clearly, while all is steady one is happy. If the values start to drop then it must indicate something has changed and this could be your device, or the way it is being

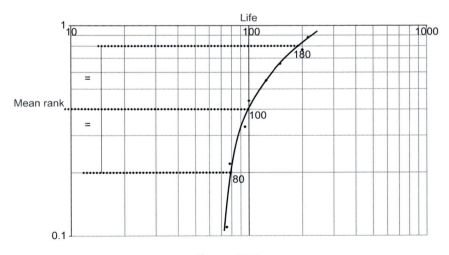

Figure 12.6
Weibull plot for data in Table 12.1(b).

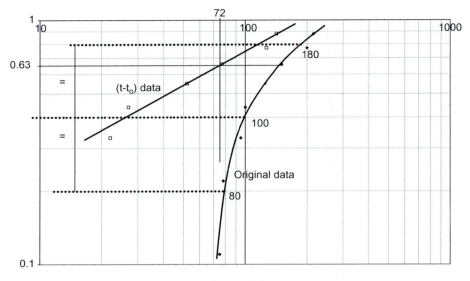

Figure 12.7
Weibull plot for modified data ($t–t_o$).

used. The other, rather sneaky, benefit is that it tells you when you need to service the device. Say that you know your device is used, on average, 10 times a month. This suggests that you should be contacting your end-user to suggest a service after 70 uses, or 7 months. A good way to keep your end-users happy is to get a sales advisor in to talk to them and avoid any breakdowns before they happen!

12.3.4 Measles Chart

The measles chart is a good way of keeping track of complaints and defects. It is so simple it is beyond comparison, but it is so powerful you will not regret beginning to use it. It is a simple sketch of the device, and every time you receive a complaint, or detect a nonconformity or defect, you plot this on the picture.

Figure 12.8 is a sketch of a device. The sketch consists of two diagrams. Figure 12.8(a) illustrates the sketch of the device when the data collection begins. It is devoid of information. However, Figure 12.8(b) shows what it looks like after several months. Every time a complaint was received an X was placed over the source of the problem. This diagram shows that there have been eight complaints – not too bad, but not too good either. The power of this diagram is that it *shows* that there is a real problem with one of the dials.

I cannot stress how powerful this diagram is. Its power comes from the simplicity of its use and then the visual power of the analysis.

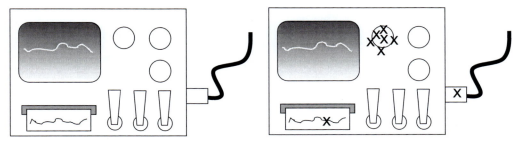

Figure 12.8
Example measles chart: (a) at the start; (b) after a few months.

Table 12.2: Example Pareto Analysis

Device (Ranked)	No. of Complaints	Cumulative Complaints	Proportion
1	55	55	0.37
2	50	105	0.7
3	25	130	0.87
4	12	142	0.95
5	5	147	0.99
6	2	149	1
	Total	149	

12.3.5 Pareto Analysis

The Pareto rule is the 80/20 rule. It states that 80% of your complaints will arise from 20% of your devices (or in the case of one device – 80% of the issues arise from 20% of the components). To perform this analysis, collect the data as individual complaints and build a table as illustrated by Table 12.2.

Figure 12.9 illustrates the data in Table 12.2 plotted on a Pareto plot. Using the 80% as a threshold demonstrates that the devices ranked as No. 1 and No. 2 require examination; this will reduce complaints significantly.

12.4 Using Your Existing Contacts

12.4.1 Early Adopters and Key Opinion Leaders

Key opinion leaders[3] (KOLs) and early adopters (EAs) are essential allies in your battle to achieve perfection. While it is ideal to produce a device that is 100% perfect at its first edition, this is often unlikely. If you have a team of KOLs and EAs available to you they can often provide excellent feedback on a potential device before it even reaches an end-user.

---

[3] A key opinion leader is someone that all other clinicians in said field would look up to. Someone who if you said "B has got one" they would be impressed.

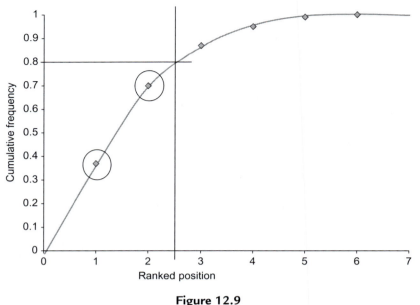

Figure 12.9
Example plot from data in Table 12.2.

However, their real benefit comes as the first tranche of users when your device is released. Rather than rushing headlong into a full launch, letting your KOLs and EAs have "first use" can reveal end-user issues while they are in your control.

They also have a further use for you – they can provide the information required to produce the key documents discussed in the previous chapter. Furthermore they are able to tell the clinical community how to use your device in a clinical environment: it is highly unlikely they would listen to you! Another of their uses is as leaders of your focus groups.

12.4.2　Focus Groups

We discussed focus groups when we discussed ideas generation. However they have further benefits and that is to provide key developmental information for your products.

Many companies use focus groups to bring new products to the end-user community before they are released. It is a good idea to do this in an environment that is as close to a clinical environment as possible (or as close as possible to the real environment in which the device is to be used). Hence if your device is to be used on a ward, try and do your initial demonstration in an environment as close to a ward as possible. Then you should get your end-users to use your device; and let them be as rough as they like. Make sure your device doesn't let you down (as it could be embarrassing) – but also don't expect them to perform somersaults; they will always find something!

Once your device is in the marketplace (fully) it is the focus group that can help you to decipher the clinical literature that is so important to your PMS.

12.4.3 Courses and Conferences

Hopefully, once your device has been released, it will gain a life of its own and end-users will start to produce their own clinical investigations using your device – often without you knowing it. The one major venue where this information will be presented is a conference. It is therefore a very good idea to keep a watchful eye on relevant conferences and their respective conference programs. If there is anything of interest you should do your best to attend, watch, and learn.

The second point of interest is that the authors of the conference papers are probably going to be very useful members of your focus group!

In addition to conferences another potential vehicle for intelligence collection is a short course. Many clinical colleges run courses on a regular basis. The colleges are always looking for subjects (and sponsors) – why not be that partner?[4] Furthermore, clinical schools in universities and colleges are also always looking for similar partnerships – again, why not be that partner? At the end of the day the students going through these programs are going to be the end-users of the future.

12.5 Vigilance

Every regulatory body expects you to have an appropriate and active system to collect and analyze complaints. This process should be able to detect complaints that are not critical to your device's safety and those that provide a risk to the end-user, the patient, or anyone else associated with it. It is beyond the scope of this book to describe how your vigilance procedure should operate; however I am duty bound to tell you that you should make sure you have one in place. As usual the FDA and MHRA websites have guidelines to help you.

12.6 The Good, the Bad, and the Ugly

The phrase "the good things in life" always passes people by. We spend so much of our lives looking for what has gone wrong that we forget to look at what has gone right. Just like little Polyanna,[5] we should learn to play the "glad" game and find the things that have gone well.

---

[4] A word of warning – take care that any partnership with training of end-users does not put you into any breach of laws regarding "undue influence," or worse, bribery. A rule of thumb to avoid this is always try and partner with at least one other company: but this alone does not give total protection.

[5] Along with *Little Women* and *The Railway Children* this is one of my wife's favorite films; she is always telling me, and the children, to play the "glad" game too.

This quirk of human nature is just the same with our products. When the medical devices regulations discuss PMS they actually force us to look in a negative direction by using terms such as *vigilance* and *complaints*. However, now that we are clever designers we can turn things to our advantage. Instead of only using the PMS structure to collect details of faults, etc., let us use it to collect details of good practice and items of excellence in our products. Doing this we not only learn from our mistakes, we learn from our successes too.

For example, your clinical end-users may be telling your sales staff how the "knob" is just right. When they are in a panic the knob is just the right size and feel to work how it is supposed to. If this is the case why not use this design for all of your other knobs? Certainly, if this piece of information is not captured then *you* will never get to know.

You have probably spotted the issue. In the event of a complaint or a hazard the end-user will contact you directly – you can count on that. However, do you think they will ring you up to tell you about how good something is? No, that's highly unlikely. Hence this sort of information only comes through your relationships with your end-users; and this normally means your sales force. You must insist that any information (good, bad, or ugly) gets back to you. With modern Internet-based communications this is easily achieved using a good contact management system (CMS) that forces the sales force to produce a report of any visit or meeting.

12.7 Summary

This chapter intended to demonstrate to you that your part in the life of a design does not end once it has been released. Rather, the work has only just begun! Several tools have been presented to help your PMS become an active design tool rather than it being a reactive tool to act on complaints. This chapter also emphasized the need for you to have an active complaints/ vigilance procedure in place – and that this is mandatory not selective.

References

Carter, A. D. S. (1986). *Mechanical reliability*. Basingstoke: Macmillan Education.

Further Reading

Bicheno, J., & Catherwood, P. (2005). *Six sigma and the quality toolbox*. Buckingham: Picsie Books.
DEC 2012., Medical evices Vigilance. MEDDEV 2.12/1 Rev 7.
FDA 2006., Post Market Surveillance under section 522 of the Federal Food, Drugs and Cosmetics Act.

## CHAPTER 13

Protecting Your IP

13.1 Introduction

Intellectual property (or IP for short) is a scary concept to most new designers. There is the constant fear that a "wolf is at the door," ready to pinch your best ideas and take them for themselves. Certainly there is a chance that someone will pinch your ideas if you do not protect them. It is the protected element of your ideas that is intellectual property; if they are unimportant you would not bother protecting them!

Why protect IP? The first and foremost reason is to ensure that all your effort put into designing a new idea or device, your costs, and all of your blood, sweat, and tears does not go to waste. The first patents were granted in about 500 BC in Ancient Greece. It was recognized that

> *"encouragement was held out to all who should discover any new refinement in luxury, the profits arising from which were secured to the inventor by patent for the space of a year."*
>
> *(Wikipedia, 2012)*

The philosophy has basically stayed the same. By being granted a patent the inventor has an effective monopoly for the life of said patent. The first recorded patent was in 1421, and concerned a floating barge with lifting gear. Presumably the inventor had developed an advantage over his competitors as he could deliver marble anywhere on the River Arno and was not bound to places where a crane had been built on the bank. The first UK patent was granted in 1449, by King Henry VI, and was concerned with the making of colored glass. This IP was actually brought into the country from outside and hence the patent protected this inventor's right to produce glass in this way – presumably it had cost him to do this. In those days international communication was a lengthy process, hence bringing a technology from somewhere else and establishing it in your own country was, literally, inventive.

The birth of the Industrial Revolution and the creation of wealth based on one's IP led to a growth in patent applications, and in the type of patent. In the USA over 150,000 patents are filed per year, and this is growing. My latest GB patent was #2,427,141 in a list that reaches all the way back to that first one in 1449. In this chapter we shall be looking at how you can join that list of inventors who have secured their IP so that you, too, have a monopoly. Indeed we shall also be looking at how to protect your IP with other ingenious devices.

Medical Device Design.
DOI: http://dx.doi.org/10.1016/B978-0-12-391942-7.00013-1

13.2 Types of IP Protection

There are four main types of IP protection:

- Patent
- Registered Design
- Trademark
- Copyright

We shall be considering the first two the most.

13.2.1 Patent

A patent is a legal document, normally drawn up by a patent attorney, which provides claims that define a new device, process, etc. and grants the owner of the patent exclusive rights to use it. Of course there is an awful legal definition – but I do not intend to belabor the point any more. However, to be eligible to apply for a patent your invention must (IPO, 2012)

- be new;
- have an inventive step that is not obvious to someone with knowledge and experience in the subject; and
- be capable of being made or used in some kind of industry.

It must not be any of the following:

- a scientific or mathematical discovery, theory, or method
- a literary, dramatic, musical, or artistic work
- a way of performing a mental act, playing a game, or doing business
- the presentation of information, or some computer programs
- an animal or plant variety
- a method of medical treatment or diagnosis
- against public policy or morality

Do not be afraid by the bullet saying medical treatment is barred. This means you cannot patent having to cut a hole in the stomach wall to access the stomach to remove a tumor; the instruments that are required to do it are not barred.

The patent is normally filed in your country of residence. However you may file in another country first if that is likely to be your first point of use. The date of this initial filing is called the *priority date*; and it is by this date that the patent lives and dies. A patent's normal lifespan is 20 years, but there are ways of extending this if the patent has lain dormant for a very good reason. If you file a patent in more than one country it is the priority date of the initial filing that matters.

Although the filing of a patent is relatively cheap, the cost of drawing up the document can, with ease, run into thousands of dollars. But this cost pales in comparison when compared with the upkeep of the patent. A modest patent portfolio that covers the major markets (i.e., EC, USA) could easily cost $40,000–$100,000 per year. Hence you need to think carefully about the costs.

13.2.1.1 Filing

A patent's life starts at the point of discussion with a patent attorney about the filing of the patent. A patent attorney is not there to decide whether your patent is any good; they only follow your instructions to file – you need to do some background work. Before you approach a patent agent you should conduct a *prior art* search. Prior art is any information that is in the public domain that relates to your patent – and it can be in any field from agriculture to zoology. The only person qualified to do this search is yourself; you know your device and you know your IP, hence you are best placed to investigate anything that has gone before. All patent offices have online search engines that enable you to look through the patent history. Indeed the UK office has a wonderful search engine at http://gb.espacenet.com/ that allows you to search global patent databases (Figure 13.1). Do not stop here: prior art need not only be patents. It can be publications, news print – anything that is quantifiably in the *public*

Bibliographic data: EP2421434 (A1) — 2012-02-29

★ In my patents list   Previous ◄   1 / 500   ► Next   ↗ Register   → Report data error   🖨 Print

IN VIVO 1H MAGNETIC RESONANCE SPECTROSCOPY FOR THE DIAGNOSIS OF TESTICULAR FUNCTION AND DISEASE

Page bookmark	EP2421434 (A1) - IN VIVO 1H MAGNETIC RESONANCE SPECTROSCOPY FOR THE DIAGNOSIS OF TESTICULAR FUNCTION AND DISEASE
Inventor(s):	TUREK PAUL [US]; KURHANEWICZ JOHN [US] ±
Applicant(s):	UNIV CALIFORNIA [US] ±
Classification:	- international: *A61B5/055*
	- european: A61B5/055; G01N33/68T; G01R33/465
Application number:	EP20100767794 20100422
Priority number(s):	WO2010US32119 20100422; US20090171758P 20090422

Abstract not available for EP2421434 (A1)

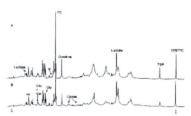

Figure 13.1
A typical result from a patent search using Espacenet using keywords Magnetic and Resonance.

domain (that is, a member of the public can access it from somewhere). It is prior art that will be the downfall of your patent if you do not conduct a proper prior art search and if you do not keep your work confidential before filing.

The critical reason for hiring a patent attorney is the development of the claims. It is the claims that carry the patent forward. The claims *must* be written by someone who knows what they are doing – to do so without this detailed knowledge is akin to performing an appendectomy on one's wife because one has watched the whole second series of *Scrubs*.

Once developed the patent agent will file your patent with the relevant office. You will receive a notification of filing and from that date on you can undertake any number of public domain events without the fear of IP theft. However, the patent application is kept secret for at least 1 year (often 18 months) – so you may wish to use this time to perfect your launch while keeping your "powder dry." During this period your device is "patent pending": you should use this 12-month window to look at any innovations you have incorporated since filing as you are able to modify the patent application (within reason) during this period.

13.2.1.2 Examination

At some point, when the backlog of applications has departed, your patent application will come to the top of the list. At this point a patent examiner will examine your application. Their job is to make sure your claims are new. This is where your stated prior art comes in; if you have done your job correctly then the patent examiner will not find any other prior art than the ones you have already cited. Because you have cited them you will have already stated what is new.

It is highly unlikely that an examiner will not produce some prior art in their examination. They *will* contest one or more of your claims, and your application will be rejected at the first examination. This is not the end – it is a part of the process. You will need to produce arguments as to why your claims are valid or, with your patent attorney, produce different claims which avoid any conflict. The patent attorney would then return this new, modified, reasoned application to the examiner. Hopefully they will see sense and grant your patent. If not another rejection may come your way. At some point the examiner will state that enough is enough and a response is not welcome; this is the end of the road for your patent and you have been unsuccessful. However, the more pleasant outcome is that the examiner states enough is enough, and they now agree that you have a new invention worthy of patenting and your patent is granted. You receive a lovely certificate (called a *letters patent*) stating that you, as the applicant, are the proud owner of a patent.

Figure 13.2 illustrates the first page of a granted patent – in this case for an orthopedics device. It reveals a great deal of information. Not only do you know the inventors, but you also know who the applicant is. The "also published as" list demonstrates that this has a USA filing, a Japan filing, an Australia filing, and the EP in the title suggests it is being filed in the EC.

SYSTEM AND METHOD FOR FRACTURE REDUCTION

Page bookmark	EP2341856 (A1) - SYSTEM AND METHOD FOR FRACTURE REDUCTION
Inventor(s):	TERRES JAYSON J [US]; AHMAD SHAHER A [US]; THORNHILL LISA R [US]; MILES III SOLON B [US]; MOCANU VIOREL NMI [US] +
Applicant(s):	OSTEOMED L P [US] +
Classification:	- international: *A61B17/80; A61B17/88*
	- european: A61B17/80A5; A61B17/80H4
Application number:	EP20090790019 20090701
Priority number(s):	WO2009US49345 20090701; US20080176677 20080721
Also published as:	☐ US2010016900 (A1) ☐ WO2010011477 (A1) ☐ JP2011528603 (A) ☐ AU2009274289 (A1) → AR074043 (A1)

Abstract not available for EP2341856 (A1)
Abstract of correspondent: US2010016900 (A1)

Translate this text into

A system for fracture reduction includes a reduction plate for reducing a fracture between a first bone segment and a second bone segment. The reduction plate includes, on a first side of the reduction plate, a travel slot and a screw hole. The travel slot is configured to slidably engage a first positioning element that extends into the first bone segment through the travel slot and the first screw hole is configured to affix the first side of the reduction plate to the first bone segment. The reduction plate includes, on a second side of the reduction plate, an adjustment hole and a second screw hole. The adjustment hole is configured to engage a second positioning element that extends into the second bone segment through the adjustment hole and the second screw hole is configured to affix the second side of the reduction plate to the second bone segment.

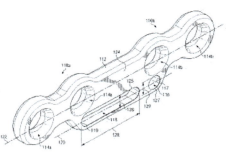

Figure 13.2
The final outcome – a granted patent.

Once granted, the patent is published and is placed in the public domain. Figure 13.3 illustrates the information available. The "description" gives an outline to the project; the "claims" lists the claims made; and the "mosaics" are IP-speak for drawings (images). Even the status of the patent is available – it may have lapsed and hence is free to use! The "cited documents" link is interesting as it is this link (and the citing documents link) that help you to develop your prior art (discussed earlier).

13.2.1.3 Other Countries

Once filed, and some time later, your patent attorney will ask you about other countries. You may be happy with having a patent in only one country (say, the USA). It is more common for medical devices to be filed in many countries. Only you will know which countries these need to be, but you should cover the countries in which you intend to sell and manufacture. Filing in other countries is an expensive business so do not file everywhere as a blanket – unless you happen to be multimillionaire.

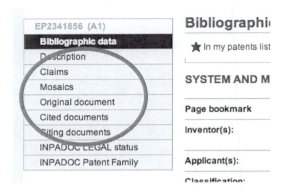

Figure 13.3
Further information *(extract from Espacenet window).*

It is worth noting that you may wish to sell your patent to a bigger player. In this case think about where they are likely to want the patent filed – almost definitely the list will include the EC, USA, and Japan. Your patent attorney/agent will be able to advise you on the current most efficient method to do this.

13.2.2 Registered Design (Design Patent)

This type of IP is concerned with appearance only. If you are unable to secure a patent on the basis of invention then this is a good backstop as, once granted, no one is allowed to make an object that looks like yours. For example, your device may look like a blue pony –and it has to look like a blue pony to work. A registered design would stop anyone making anything that resembled your blue pony. It does not stop them making a pink one or a blue horse.

If the appearance of your device is a big issue to your device then a registered design may be a useful method of securing some IP protection.

Figure 13.4 was produced using a search in the UK registered design database for medical and laboratory equipment/endoscope. A registered design is a very cheap way of protecting a device[1], but all your competitors need do is change a simple line or color and the protection has gone.

13.3 Keeping Mum

Arguably one of the best known sayings from the Second World War allied propaganda machine was:

"Loose lips might sink ships."

---

[1] In a recent radio interview James Dyson (of the famous Dyson vacuum cleaner) stated that his registered designs were his most valuable pieces of IP. This was, he said, because the first thing someone will copy is how your device looks...if only to fool the customer into thinking it is the "real thing," or "it looks the same, hence it must be the same."

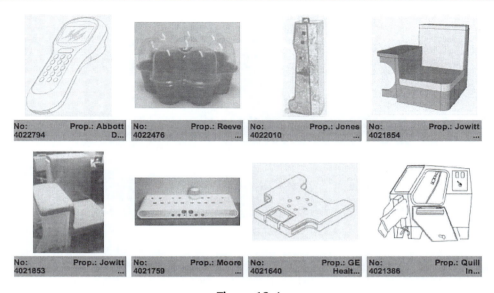

| No: 4022794 | Prop.: Abbott D... | No: 4022476 | Prop.: Reeve ... | No: 4022010 | Prop.: Jones ... | No: 4021854 | Prop.: Jowitt ... |

| No: 4021853 | Prop.: Jowitt ... | No: 4021759 | Prop.: Moore ... | No: 4021640 | Prop.: GE Healt... | No: 4021386 | Prop.: Quill In... |

Figure 13.4
Example registered designs (endoscope).

The problem with IP is that it is always visible somewhere. Until you have a patent nothing is safe. Keeping things secret is paramount. If you have a nice building all to yourself and you work on your own then keeping things secret is relatively easy. But there's always the potential to let the cat out of the bag unintentionally. So keep the "loose lips…" quote to mind (Figure 13.5).

Working with universities brings this problem to the forefront of your mind. Academics and professors live by publishing; it is their right to publish freely. If you work with an academic institution have a contract stating that all publications relative to your work must be vetted by you first. You must be able to veto anything that can go into the public domain that could jeopardize your IP.

Working with manufacturers for prototypes can also be a problem. "The tank" anecdote is a good solution here. In the First World War the British were building the first armored vehicles – they had no name. To keep them secret they had separate items made all over the country and they were then assembled in one high security plant. The manufacturers wanted to know what they were making so the military told them they were making powered water carriers – machines that moved water in large tanks. This name stuck and they have been called *tanks* ever since. But at that time who would have known what was being made? The car industry does this by using names such as "project X33Y" – again, who knows what that is? The lesson here is do not give your secret projects an identifiable name. If you are worried about someone seeing the whole, get bits made in a variety of establishments so only you see the final assembly.

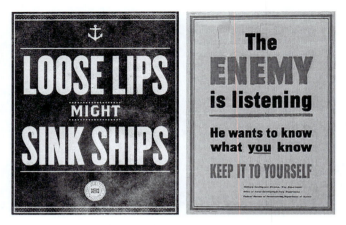

Figure 13.5
Two Second World War Allied propaganda posters.

Some companies do not file patents. It is cheaper for them to keep the invention secret. KFC and Coca-Cola are very good examples of this method of IP protection.

13.4 Talking with Partners

At some stage, before any patent filing, you will need to talk with potential partners. These may be prototype manufacturers, or they may be a university, but you will need to share confidential information with them. In this situation you should get all parties to sign a non-disclosure agreement (NDA). An NDA is a legal document barring the partners from disclosing the other partners' confidential information for a specific period of time (normally 5 years). This document allows you to enter into discussions with a partner without losing your IP.

NDAs are very difficult to formulate. There are samples available on the web – you may have already signed one and hence have a copy. In all instances get a lawyer to draw one up for your company, but do not let them make it too long! Be prepared to sign the NDA long before you meet your potential partner. They will want their lawyer to look at it to make sure they are not being obliged to something they cannot uphold.

13.5 Summary

This brief chapter has introduced you to the concept of IP protection. There is no particular advice a book like this can give you; all it can do is suggest you look seriously at IP protection. The filing of IP needs to be a business decision; if you are not going to exploit the IP why file it? It is a very expensive way of getting your name on the web! We also looked at secrecy. Once again, all a book like this can do is point out that you should be keeping things as secret as possible – and your main tool is the NDA.

If granted, your patent can be a very lucrative commodity.

References

Intellectual Property Office (2012). <www.ipo.gov.uk> cited February 2012.
Wikipedia (2012). *Patent*. <http://en.wikipedia.org/wiki/Patent> cited February 2012.

## CHAPTER 14

Obtaining Regulatory Approval to Market

14.1 Introduction

I can hear your sighs of relief! At last we have arrived at the zenith: the time of regulatory approval. Just as a reminder, in the EC this is obtaining a CE mark; in the USA this is obtaining an FDA clearance to market. In other countries, such as India, Australia, Canada, and Japan for example, they have their own terminology. However, in all cases you are not allowed to sell your device in said country unless you have the clearance/approval to do so. In this chapter we shall concentrate on the EC (using the UK as the base) and USA application processes. Once you have understood these, the extension to any remaining countries is a small conceptual leap.

14.2 Class I Devices

In previous chapters we saw the classification process. The obvious starting point for all applications is the confirmation of the device's classification. The easiest classification of all is Class I, or in FDA speak "510(k) exempt." All manufacturers of medical devices need to be registered, and this is the sole requirement for both the EC and the FDA.

14.2.1 EC Application

As stated earlier, this is going to be based around applying using the UK as the administrative base – if you were to use Germany, for example, the process is similar but the forms will be different. The first port of call is to identify the competent authority in the relevant country; in the UK this is the MHRA (see Appendix A for other EC countries). The MHRA has a guidance document for Class I manufacturers (MHRA, 2006) – they also have a guidance document for the actual registration process (MHRA, 2008). These will help you to identify your commitments.

The form you will need is called *"Medical Devices Regulations 2002: Regulations 19 and 30 form RG2."* The form is simple to complete. There is only one form per company: it is *not* a case of one form per product! Before you start you will need to download the document called *"Appendix A and B"* (MHRA, 2008).

This document contains a list of all items *not* considered to be medical devices. It lists existing medical device groups and allocates them a Class I Generic Family Group Code. For your benefit I have summarized the codes in Appendix D; however ensure you download the latest guidance as these codes may change.

Medical Device Design.
DOI: http://dx.doi.org/10.1016/B978-0-12-391942-7.00014-3

CASE STUDY 14.1

A company wishes to sell its new peak flow meters in the EC. Identify the generic code it should adopt.

Appendix D gives this as a Diagnostic Device (Letter C) with a specific code C3.

CASE STUDY 14.2

A new company has developed a new device that requires a packed set of dressings to be used in the operating theater. Identify the generic code.

Appendix D illustrates two answers. If the dressings are individual then their code would be D3. If, however, the dressings are supplied as a pack for a particular procedure then this could fall under code L2. The obvious answer is to register for both, but note that you must comply with any further regulations required for System and Procedure Packs.

CASE STUDY 14.3

An established UK company is to act as the representative for a USA-based company in the EC. Identify the information required for form RG2.

The authorized representative's name and address is entered into the "UK Address" section in Part 2. The USA manufacturer's name and address is entered into the "Manufacturer's Address if outside the EC" in Part 2. Note that PO Box numbers are not allowed.

The USA company's product range is listed as Codes and Product Names in Part 4.

The process for registration is similar in all EC countries; you just need to decide which country is to be your "home base." More often than not that decision is driven by the following:

- Market dominance
- Native language
- Your EC representative's address

For more information, the competent authority's contact details are given in Appendix A.

Once your registration form is complete you send it to the competent authority, with fee enclosed, and await the reply. You should receive a confirmation letter with your company registration number within a few months. This confirmation letter is your approval to market your registered devices in the whole of the EC.

Some of you may be asking, "If this is all that is required, why did we need to go through the previous 13 chapters?" Even though you have self-certified, you have actually self-certified that *you have used the previous 13 chapters in order to ensure you have a safe, usable medical device*. You can imagine the repercussions if the competent authority finds otherwise!

14.2.2 FDA Registration

Here the process is much the same. If your product is Class I it is very likely to be 510(k) exempt. Even some Class II devices can be 510(k) exempt so your classification process, discussed earlier, must be comprehensive – if only to save you time and money. In this case you need to "register" your company and "list" your devices. Note that, as with the EC process, this does not exempt you from following proper design control procedures. On the contrary, the FDA will take a very dim view if you do not.

If your company is based outside of the USA you must have a designated office or agent with a U.S. address. You must have an agreed protocol with this addressee; it cannot be a PO box number!

The first port of call is to register your establishment. To do this you must have an account. The process is quite protracted but "How to register and list" in the medical device sections on the FDA website is useful. You will need to decide what sort of establishment you are (FDA, 2012):

1. Contract Manufacturer – Manufactures a finished device to another establishment's specifications.
2. Contract Sterilizer – Provides a sterilization service for another establishment's devices.
3. Foreign Exporter – Exports or offers for export to the United States (U.S.), a device manufactured or processed by another individual, partnership, corporation or association in a foreign country, including devices originally manufactured in the United States. A foreign exporter must have an establishment address outside the U.S.
4. Initial Distributor – Takes first title to devices imported into the U.S. An Initial Distributor must have a U.S. address.
5. Manufacturer – Makes by chemical, physical, biological, or other procedures, any article that meets the definition of "device" in Section 201(h) of the Federal Food, Drug, and Cosmetic (FD&C) Act.
6. Repackager – Packages finished devices from bulk or repackages devices made by a manufacturer into different containers (excluding shipping containers).
7. Relabeler – Changes the content of the labeling from that supplied from the original manufacturer for distribution under the establishment's own name. A relabeler does not include establishments that do not change the original labeling but merely add their own name.
8. Remanufacturer – Any person who processes, conditions, renovates, repackages, restores, or does any other act to a finished device that significantly changes the finished device's performance or safety specifications, or intended use.

9. Reprocessor of Single Use Devices – Performs remanufacturing operations on a single use device.

10. Specification Developer – Develops specifications for a device that is distributed under the establishment's own name but performs no manufacturing. This includes establishments that, in addition to developing specifications, also arrange for the manufacturing of devices labeled with another establishment's name by a contract manufacturer.

11. U. S. manufacturer of export only devices – Manufactures medical devices that are not sold in the U.S. and are manufactured solely for export to foreign countries.

For the purposes of this exercise your company would be likely to fall into one of two categories: #5 – Manufacturer or #10 – Specification Developer. The FDA discriminates between a company that both "designs and makes" and one that "designs only but has others who make it for them."

For the device(s) you wish to list you must know the product code (see classification rules in previous chapters) and the regulation number. You can find the specific codes related to your device on the FDA website; here you can download all of the product codes for all devices. For example, Table 14.1 illustrates the codes for a retractor.

The first column is the review panel. The second column is the specific area, called the medical specialty (described in more detail in Table 14.2). The product code is unique to this generic device; the regulation number is not specific to a device but to a family of devices, as the further two rows demonstrate.

All applications are electronic. You must have created a company account (that is, registering your company with the FDA); once registered you will be given a unique account and then you can start listing your devices.

Unlike the EC this is an annual process and you will need to pay an annual registration and listing fee. Do not forget to update your registration every year (normally by 31st December).

Table 14.1: Example Codes for a Retractor

Review Panel	Medical Specialty	Product Code	Device Name	Regulation Number
DE	DE	EIG	Retractor, All Types	872.4565
DE	DE	EIK	Carver, Wax, Dental	872.4565
DE	DE	EIL	Gauge, Depth, Instrument, Dental	872.4565

Table 14.2: Medical Specialty Codes

Medical Specialty (Advisory Committee)	Regulation No.	Medical Specialty Code
Anesthesiology	Part 868	AN
Cardiovascular	Part 870	CV
Clinical Chemistry	Part 862	CH
Dental	Part 872	DE
Ear, Nose, & Throat	Part 874	EN
Gastroenterology & Urology	Part 876	GU
General Hospital	Part 880	HO
Hematology	Part 864	HE
Immunology	Part 866	IM
Microbiology	Part 866	MI
Neurology	Part 882	NE
Obstetrics/Gynecology	Part 884	OB
Ophthalmic	Part 886	OP
Orthopedic	Part 888	OR
Pathology	Part 864	PA
Physical Medicine	Part 890	PM
Radiology	Part 892	RA
General & Plastic Surgery	Part 878	SU
Clinical Toxicology	Part 862	TX

As with the EC registration, this is self-certification – you are certifying that you have done everything required. More fool you if you have not. FDA inspectors are Federal Marshals: they are not to be treated lightly!

14.3 Higher Classifications

Unfortunately the processes of the USA and the EC are now as different as chalk and cheese. Neither is a precursor to the other. Success in one does not automatically lead to success in the other. Both systems are wholly different. However, the saving grace is that the information you need for both is virtually identical: it is the way it is presented that differs. The FDA process is a paperwork process. The EC process is an audit-based process.

14.4 FDA Process

The 510(k) process is, arguably, easier to undertake: hence we shall look at this first. You will need to register your establishment as described in the previous section. Nothing can happen until that expense has been undertaken. However, before you progress further you should confirm if your company is classed as a small company in FDA eyes. If so, most of the future application fees are much reduced. Hence it is worth undertaking a quick trip onto the FDA website and seeking out the most current rules.

The 510(k) submission is now wholly electronic, but you will need to produce a submission document. It is a good idea to build this in electronic form, then print it and put it in a ring binder. This simple process ensures two things:

1. You do not leave any section blank.
2. You use U.S. letter size paper: *not* A4!

The document has standard sections that you can download from the FDA website. You should download the Format for Traditional and Abbreviated 510(k) submissions guidance document (FDA, 2005) and have this close at hand. Furthermore, the FDA website contains very detailed instructions on their website.

Your submission document must contain all 21 sections (Table 14.3). If you have no entry in a section, do not remove it. Put in a statement that states that this section is not relevant and give a reason. For example, "This device does not contain software."

Most sections have a specific format that the FDA is looking for. Hence there are two pieces of advice:

a. Use the FDA guidance sheets as they tell you exactly what the FDA examiners want and in what format they want it.

Table 14.3: 510(k) Submission Sections

1. Medical Device User Fee Cover Sheet (Form FDA 3601)
2. CDRH Premarket Review Submission Cover Sheet
3. 510(k) Cover Letter
4. Indications for Use Statement
5. 510(k) Summary or 510(k) Statement
6. Truthful and Accuracy Statement
7. Class III Summary and Certification
8. Financial Certification or Disclosure Statement
9. Declarations of Conformity and Summary Reports
10. Executive Summary
11. Device Description
12. Substantial Equivalence Discussion
13. Proposed Labeling
14. Sterilization and Shelf Life
15. Biocompatibility
16. Software
17. Electromagnetic Compatibility and Electrical Safety
18. Performance Testing – Bench
19. Performance Testing – Animal
20. Performance Testing – Clinical
21. Other

b. Use an example of good practice from another medical device manufacturer; you only need ask and a refusal should not offend.

And another suggestion:

c. The FDA examiners are extremely helpful, so why not ask them?

14.4.1 Substantial Equivalence (SE)

Section 12 is very important. This is the section where you claim "substantial equivalence": your device is so similar to other devices in the market that it is – in effect – the same. This simple section helps tremendously. But to do this you must find precedents. This is done using the FDA 510(k) search engine (in a similar vein to that shown in previous chapters).

For example, let us suppose your company wishes to register a c-arm image intensifier. The 510(k) search results are shown in Figures 14.1 and 14.2.

The results of your 510(k) search will help you with your submission. In Figure 14.2, the item circled #1 is the 510(k) number you *must* refer to in your SE discussion; #2 identifies the relevant FDA codes (again, you must refer to these for any specific standards you have used); #3

510(K) Premarket Notification
○ FDA Home ○ Medical Devices ○ Databases

1 to 10 of 71 Results for c-arm

| 1 | 2 | 3 | 4 | 5 | 6 | 7 | 8 | > |

10 ÷ results per page

New Search

Export To Excel Help

Device Name	Applicant	510(K) Number	Decision Date
Series 7700 Mobile C-Arm, Compact 7700 Mobile C-Arm, Compact 7700 Plus Mobile C-Arm	GE DEC MEDICAL SYSTEMS	K000221	04/11/2000
Apollo Mobile "C" Arm System	CARES BUILT, INC.	K010393	03/12/2001
Isi-2500 Ccd C-Arm, Isi-2500 Plus Ccd C-Arm	IMAGING SERVICES, INC.	K010772	08/23/2001
Moonray Mobile C-Arm	SIMAD S.R.L.	K013426	03/26/2002
Stealthstation System Three Dimensional C-Arm Interface	MEDTRONIC SURGICAL NAVIGATION TECHNOLOGIES	K022414	08/14/2002
Model Ami1200 C-Arm	INTEGRITY PRACTICE MANAGEMENT, INC.	K022911	12/02/2002
Kmc-950 C-Arm Mobile System	UNITED RADIOLOGY SYSTEMS, INC.	K032761	05/14/2004
Lateral Angiographic C-Arm Support Mh-400	SHIMADZU CORP.	K033184	12/09/2003
Shimadzu Surgical Mobile C-Arm Imaging X-Ray System, Model Wha-200	SHIMADZU CORP.	K043379	02/17/2005
Centrion 500 C-Arm System	OSTEOSYS CO., LTD.	K050866	04/27/2005

Figure 14.1
510(k) search result for c-arm.

510(k) Premarket Notification
◎ FDA Home ◎ Medical Devices ◎ Databases

510(k) | Registration & Listing | Adverse Events | Recalls | PMA | Classification | Standards
CFR Title 21 | Radiation-Emitting Products | X-Ray Assembler | Medsun Reports | CLIA | TPLC

New Search   Back To Search Results

Device Classification Name	System, X-Ray, Fluoroscopic, Image-Intensified
510(K) Number 1.	K010772
Device Name	ISI-2500 CCD C-ARM, ISI-2500 PLUS CCD C-ARM
Applicant	IMAGING SERVICES, INC.
	8210 Lankershim Blvd., #1
	North Holloywood, CA 91605
Contact	Dean James
Regulation Number 2.	892.1650
Classification Product Code	JAA
Subsequent Product Code	OXO
Date Received	03/14/2001
Decision Date 3.	08/23/2001
Decision	Substantially Equivalent (SE)
Classification Advisory Committee	Radiology
Review Advisory Committee	Radiology
Summary 4.	Summary
Type	Traditional
Reviewed By Third Party	No
Expedited Review	No
Combination Product	No

Figure 14.2
Detail from 510(k) search.

states that this application was based on an SE, hence it's summary – #4 – will not only supply you with a format for the summary but pointers to even more devices with which to claim SE.

14.4.2 Other Sections

All other sections are self-explanatory. As stated previously, the FDA examiners give advice and their website contains substantial step-by-step instructions.

14.4.3 Process

Once you have submitted your 510(k) application you may wait for months before receiving a formal examination report. Certainly if you have not stuck to standard format you will be waiting a long time! Do not be surprised if you receive instructions requiring your submission to be modified, this is to be expected. The simplest way to remedy any failings is to ask the examiner for clarification and take it from there. Do not get angry, do not suggest they know not what they do; this does not help anyone. Instead simply ask them how you can meet their requirements: this will always reap rewards!

Do not expect this process to be quick. The higher the classification, the more detailed the inspection. The process can take a few weeks from receipt to several months, even years.

Eventually, and with luck, you will receive a confirmation letter that you have clearance to market (*not* approval). This letter will contain your 510(k) registration number and the registration will be added to your company's electronic record. Remember to renew your FDA registration every year or your 510(k) lapses.

14.4.4 Effect on IP

The USA has recognized the impact of the regulatory process on the life of a patent. Hence it is possible to claim for a PTE (Patent Term Extension). But this will be dependent on your patent agent successfully claiming the regulatory process has impacted on your patent's life. The duration of the extension is also dependent on your patent agent's arguments. Unfortunately there is no such provision in the EC for medical devices.

14.5 EC Process

Unfortunately the EC process is more difficult as it is based on an audit of your procedures. Unlike the FDA where you need to prove the device, the EC process is to prove that you have procedures in place that enable you to design, make, and sell a safe medical device, and that you adhere to them.

The process is to identify a notified body (the next level down from the competent authority) who has a license to audit a company to EC 93/42/EC and/or ISO 13485. In order to be able to attach a CE mark to your device, you must have the 93/42/EC certificate. There is a plethora of accreditation bodies that can perform this task and their costs and helpfulness vary. The best advice here is to ask around and find out what other establishments recommend: there is no more credit to obtaining a certificate from an expensive group – it is still a certificate.

The audit itself will take 1–2 days. This will be the most intense two days you will have all year. However, be prepared to spend the preceding week(s) preparing for the visit. When they arrive do not be gladiatorial; this does not help. However, do be prepared to stand your ground if you feel it necessary as they are *not* experts in every field of medicine.

14.5.1 Advice

Unlike the FDA, the body you ask to audit your company is *not allowed* to give advice; this is against the rules. It is therefore necessary to seek advice, from an early stage, from an "independent advisor." The costs and level of expertise vary with the tides. Once again, ask around and interview advisors. It is strongly suggested that unless you are a highly experienced medical devices auditor, you get an advisor as soon as possible and let them "hold your hand" through the whole process. The initial costs may hurt, but the savings they will make in the long run will pay you back many times over.

14.5.2 Documentation

ISO 13485 is a quality standard, so do not be surprised if you are asked to produce your quality manual and procedures. Even if you only go for the CE certificate alone you will still require these. You may be only CE marking the design of a device, but the auditors will still want to audit the whole process from design to manufacture; from buying stock to making individual lots; from writing the IFU to translating it into German; from sales to postmarket surveillance; from complaint to vigilance. Once again obtaining good advice at the start will help.

How you lay out your documentation is down to your own corporate systems, but some you cannot avoid; here are a few to consider.

14.5.3 Technical Files (DHF)

It is with a high degree of certainty that I predict that your device's technical files will be examined. Do not expect them to query your design – they will check that you have followed your design procedures. They will check things such as inputs, risk analysis, design changes, and the clinical evaluation report. In general the auditors will not be interested in the technical files of Class I, but do not let this fool you – they could ask for one to see if you are really applying your design procedures across the board.

14.5.4 Standards

You should have a register of standards and regulatory documents. All should be up to date and your process should demonstrate that they have been audited for relevance and status at least annually. Most certainly you should have copies of the MDD (all versions), ISO 13485 and ISO 14971 at hand. Also, do not forget the new Clinical Evaluation for Medical Devices standard and all others that you refer to in your documentation.

14.5.5 Stock Control

You must have a stock control procedure in place. You must have complete traceability of all your devices. If someone rings you to ask for the material specifications and source for Lot No. 1245X you must be able to find this information. Equally, if your material supplier says you need to find all components made from their Lot v345 you must be able to do so. The auditors will check this with rigor.

14.5.6 Advice

For the first time "medical devicee" there is little to suggest apart from seek advice. Unfortunately no number of short courses makes up for the experience of being audited. Seek out similar companies to yours and ask advice. Find a good consultant to walk you through the process: do not let them write everything for you but get them to help you write your own. Also, ask them to be with you during the audit process. This simple request will repay itself very quickly.

14.5.7 The Outcome

After your audit do not be surprised to get comments and suggestions for improvement. These will come as minor or major nonconformities. Normally a minor nonconformity is something small and non-life threatening, hence the need for a rapid response is not demanding. A major nonconformity, however, can lead to no CE mark: this means something serious is wrong and needs to be put right. It does not mean you have failed, it just means there is something wrong; you should treat all nonconformances as learning opportunities (remember from the ashes of disaster…). We have all seen major nonconformances in our time, and we have all had to deal with them; but we have all improved our processes as a result.

You will need to agree on a scope with your auditor. The scope is what appears on your certificate so if you sell x-ray machines a scope of kidney dishes is useless. Equally do not be so proscriptive that it leaves you with no room for diversity: do not have one saying 6 mm bone screws instead of simply bone screws.

At some stage, hopefully immediately after the audit, you will receive confirmation that you have been successful and you will be assigned the relevant certificate. This is not the end; it is commonplace for the notified body to request electronic copies of the technical files (normally one per year) for inspection. Furthermore, just like the FDA, this is an annual process and you must expect to go through it again 12 months later.

14.6 Getting to Market

Unfortunately obtaining the 510(k) and the CE mark does not, in itself, guarantee sales. It only gives us the ability *to sell*. There is no doubt that the medical devices market is the most lucrative of markets; but it is also, without doubt, the hardest to penetrate. It is beyond the scope of this text to start you on a marketing strategy course but it is pertinent to look at how the design process has helped.

14.6.1 Unique Selling Points (USP)

All sales staff like an identifiable USP. What makes your device stand out from the crowd? Your PDS should have identified improvements or design differences. The most commonly wanted ones are those derived from customer input, e.g., it is easier to use, or it fits in the palm of your hand. If you conducted the House of Quality analysis you will have compared your device against your competitors and hence identified USPs.

The clinical evaluation report should have identified improvements on previous devices. Your device may have shortened the procedure by 50%; it may have resulted in a device that is 50% more reliable. All are valuable marketing commodities.

14.6.2 Key Opinion Leaders (KOLs)

All medical devices are in fields of medicine where there is a key opinion leader. This is the one (or more) person who everyone says "Has X got one?" and if you answer "yes" they then reach for their wallet. While this is not guaranteed, without the answer "yes" everything becomes more difficult.

You should seek out the KOL(s) in your area and introduce them to your device. Having them in at the front of the process (i.e., in the focus group) will help tremendously.

Do not always assume that a KOL needs to be clinical. In the USA the KOL may be a medical insurance company. In the UK it may be the National Institute for Health Clinical Excellence (NICE). The definition of a KOL is very broad.

14.6.3 Independent Studies

Unfortunately it is a truism to say that any studies you have conducted will be tainted with the grain of commercial bias. It is hard to prove that one has no bias towards one's own baby so it is far better to get someone else to do this for you. All teaching hospitals have registrars (interns) looking for research projects. Your key opinion leaders will also want to conduct studies. The power of having a paper written by someone not associated with the product cannot be underestimated: they are worth their weight in gold. Furthermore, when completed they update your clinical evaluation report!

14.6.4 Health Economics

There is little doubt that your device will fall into one of two camps. It will either be cheaper than the rest (hence it is a no-brainer sale); or it will be cost neutral or slightly more expensive. In the latter case you will need to persuade the hospital that your device saves them money in the long run. This could be due to lower running costs, or less medication required. Either way the QALY analysis and clinical evaluation report, discussed previously, will provide all the information you require.

14.6.5 Insurance

Do not forget insurance! The NHS, for example, demands £5 million cover as a minimum just to get on their register. Before you sell anything you need to get adequate cover.

Once again ask around. You will not be able to obtain medical devices cover from your high street insurance broker: it is a specialist field. Asking questions should never cause embarrassment.

14.7 Summary

We have met the self-certification process of the EC and the FDA for Class I and 510(k) exempt devices. We saw that the declaration we make means that we cannot avoid having a controlled design process – and that to do otherwise is highly risky. We saw that this was the simplest route to market but is limited to the most simple of medical devices.

We subsequently saw that the FDA and EC process diverts for higher classification devices and we were introduced to the 510(k) application process and the CE mark audit process. We recognized the need for advisors, especially when applying for the first time, and we also saw that it was an annual process.

We also saw how our design process has helped us to help our sales staff. We will have developed USPs, found key opinion leaders, and produced evidence that they can use to persuade the most hardened of purchasing officers.

References

FDA (2012). *Who must register, list and pay the fee.* <www.fda.gov>.
FDA (2005). *Guidance for industry and FDA staff: Format for traditional and abbreviated 510(k)s.* FDA.
MHRA (2006). *Guidance notes for manufacturers of Class I medical devices, no 7.* MHRA.
MHRA (2008). *Guidance notes for the registration of persons responsible for placing devices on the market, no 8.* MHRA.
MHRA (2008). *Appendix A and B.* MHRA.

Useful Websites

FDA Medical Devices

www.fda.gov/cdrh

Table A.1: European Competent Authorities for Medical Devices

EC Country	Competent Authority	Website
Austria	Österreichische Agentur für Gesundheit und Ernährungssicherheit GmbH	www.basg.at
Belgium	Institut Inspektionen, Medizinprodukte und Hämovigilanz – AGES PharmMed (AIMDD, MDD) Federal Public Service Health, Food Chain Safety and Environment Directorate General Public Health Protection: Medicinal Products Medical Devices Unit (IVDMD) Scientific Institute of Public Health Department of Clinical Biology	www.fagg.be www.wiv-isp.be
Bulgaria	Bulgarian Drug Agency Department of Medical Devices	www.bda.bg
Cyprus	Cyprus Medical Devices Competent Authority	www.mphs.moh.gov.cy
Czech Republic	State Institute for Drug Control	www.sukl.cz
Denmark	The Danish Medicines Agency	www.medicaldevices.dk
Estonia	Health Board	www.terviseamet.ee
Finland	Medical Devices Department Valvira – National Supervisory Authority for Welfare and Health	www.valvira.fi
France	Agence française de sécurité sanitaire des produits de santé (AFSSAPS)	www.afssaps.fr
Germany	Federal Institute for Drugs and Medical Devices	www.bfarm.de
Greece	National Organization for Medicines	www.eof.gr
Hungary	Authority for Medical Devices Budapest	www.eekh.hu
Iceland (EFTA)	Directorate of Health	www.landlaeknir.is
Ireland	Irish Medicines Board – Medical Devices Department	www.imb.ie
Italy	Ministry of Labour, Health and Social Affairs Department of Innovation Directorate General of Medicine and Medical Devices	www.sanita.it

Medical Device Design.
DOI: http://dx.doi.org/10.1016/B978-0-12-391942-7.00024-6

Table A.1: European Competent Authorities for Medical Devices

EC Country	Competent Authority	Website
Latvia	State Agency of Medicines Medical Devices Evaluation Department for Latvia	www.zva.gov.lv
Liechenstein (EFTA)	Amt für Gesundheit	www.llv.li
Lithuania	The State Health Care Accreditation Agency under the Ministry of Health of the Republic of Lithuania	www.vaspvt.gov.lt
Luxembourg	Ministère de la Santé – Direction de la Santé	www.ms.etat.lu
Malta	Malta Standards Authority Regulatory Affairs Directorate	www.msa.org.mt
Netherlands	Dutch Healthcare Inspectorate	www.igz.nl
Norway (EFTA)	Sosial-og helsedirektoratet Norwegian Directorate for Health and Social Affairs	www.shdir.no
Poland	The Office for Registration of Medicinal Products Medical Devices and Biocidal Products	www.urpl.gov.pl
Portugal	Infarmed National Authority of Medicines and Health Products, IP Unidade de Vigilância de Produtos de Saúde	www.infarmed.pt
Romania	Ministry of Health	www.ms.ro
Slovenia	Agency for Medicinal Products and Medical Devices of the Republic of Slovenia	www.jazmp.si
Slovakia	State Institute for Drug Control Medical Devices Section	www.sukl.sk
Spain	Ministerio Sanidad y Consumo Agencia Española de Medicamentos y Productos Sanitarios	www.aemps.es
Sweden	Medical Products Agency "Läkemedelsverket" Medical Devices	www.lakemedelsverket.se
Switzerland (EFTA)	Swissmedic Medical Devices Division	www.swissmedic.ch/md.asp
Turkey (Candidate)	Ministry of Health DG for Pharmaceuticals and Pharmacy Department of Medical Device Services Market Surveillance Section	www.iegm.gov.tr
United Kingdom	Medicines & Healthcare products Regulatory Agency (MHRA)	www.mhra.gov.uk

Table A.2: Summary of Websites Referred to in Chapter 2

American Society for Testing and Material (ASTM)	http://www.astm.org
British Standards Institute (BSI)	http://shop.bsigroup.com
FDA: Databases	http://www.fda.gov/MedicalDevices/ DeviceRegulationandGuidance/Databases/
FDA: Recognized Consensus Standards	http://www.accessdata.fda.gov/scripts/cdrh/cfdocs/ cfStandards/search.cfm
International Organization for Standardization (ISO)	http://www.iso.org

Table B.1: Normal Distribution Table for 2^k Factorial Experiment Analysis – $z(\Phi)$

(a) Positive z ($p > 0.5$)									
Z	0	0.01	0.02	0.03	0.04	0.05	0.06	0.07	0.08	0.09
0	0.5	0.504	0.508	0.512	0.516	0.5199	0.5239	0.5279	0.5319	0.5359
0.1	0.5398	0.5438	0.5478	0.5517	0.5557	0.5596	0.5636	0.5675	0.5714	0.5753
0.2	0.5793	0.5832	0.5871	0.591	0.5948	0.5987	0.6026	0.6064	0.6103	0.6141
0.3	0.6179	0.6217	0.6255	0.6293	0.6331	0.6368	0.6406	0.6443	0.648	0.6517
0.4	0.6554	0.6591	0.6628	0.6664	0.67	0.6736	0.6772	0.6808	0.6844	0.6879
0.5	0.6915	0.695	0.6985	0.7019	0.7054	0.7088	0.7123	0.7157	0.719	0.7224
0.6	0.7257	0.7291	0.7324	0.7357	0.7389	0.7422	0.7454	0.7486	0.7517	0.7549
0.7	0.758	0.7611	0.7642	0.7673	0.7704	0.7734	0.7764	0.7794	0.7823	0.7852
0.8	0.7881	0.791	0.7939	0.7967	0.7995	0.8023	0.8051	0.8078	0.8106	0.8133
0.9	0.8159	0.8186	0.8212	0.8238	0.8264	0.8289	0.8315	0.834	0.8365	0.8389
1	0.8413	0.8438	0.8461	0.8485	0.8508	0.8531	0.8554	0.8577	0.8599	0.8621
1.1	0.8643	0.8665	0.8686	0.8708	0.8729	0.8749	0.877	0.879	0.881	0.883
1.2	0.8849	0.8869	0.8888	0.8907	0.8925	0.8944	0.8962	0.898	0.8997	0.9015
1.3	0.9032	0.9049	0.9066	0.9082	0.9099	0.9115	0.9131	0.9147	0.9162	0.9177
1.4	0.9192	0.9207	0.9222	0.9236	0.9251	0.9265	0.9279	0.9292	0.9306	0.9319
1.5	0.9332	0.9345	0.9357	0.937	0.9382	0.9394	0.9406	0.9418	0.9429	0.9441
1.6	0.9452	0.9463	0.9474	0.9484	0.9495	0.9505	0.9515	0.9525	0.9535	0.9545
1.7	0.9554	0.9564	0.9573	0.9582	0.9591	0.9599	0.9608	0.9616	0.9625	0.9633
1.8	0.9641	0.9649	0.9656	0.9664	0.9671	0.9678	0.9686	0.9693	0.9699	0.9706
1.9	0.9713	0.9719	0.9726	0.9732	0.9738	0.9744	0.975	0.9756	0.9761	0.9767
2	0.9772	0.9778	0.9783	0.9788	0.9793	0.9798	0.9803	0.9808	0.9812	0.9817
2.1	0.9821	0.9826	0.983	0.9834	0.9838	0.9842	0.9846	0.985	0.9854	0.9857
2.2	0.9861	0.9864	0.9868	0.9871	0.9875	0.9878	0.9881	0.9884	0.9887	0.989
2.3	0.9893	0.9896	0.9898	0.9901	0.9904	0.9906	0.9909	0.9911	0.9913	0.9916
2.4	0.9918	0.992	0.9922	0.9925	0.9927	0.9929	0.9931	0.9932	0.9934	0.9936
2.5	0.9938	0.994	0.9941	0.9943	0.9945	0.9946	0.9948	0.9949	0.9951	0.9952

Medical Device Design.
DOI: http://dx.doi.org/10.1016/B978-0-12-391942-7.00025-8

Table B.1: Normal Distribution Table for 2^k Factorial Experiment Analysis – $z(\Phi)$
(Continued)

Z	0	−0.01	−0.02	−0.03	−0.04	−0.05	−0.06	−0.07	−0.08	−0.09
0	0.5	0.496	0.492	0.488	0.484	0.4801	0.4761	0.4721	0.4681	0.4641
−0.1	0.4602	0.4562	0.4522	0.4483	0.4443	0.4404	0.4364	0.4325	0.4286	0.4247
−0.2	0.4207	0.4168	0.4129	0.409	0.4052	0.4013	0.3974	0.3936	0.3897	0.3859
−0.3	0.3821	0.3783	0.3745	0.3707	0.3669	0.3632	0.3594	0.3557	0.352	0.3483
−0.4	0.3446	0.3409	0.3372	0.3336	0.33	0.3264	0.3228	0.3192	0.3156	0.3121
−0.5	0.3085	0.305	0.3015	0.2981	0.2946	0.2912	0.2877	0.2843	0.281	0.2776
−0.6	0.2743	0.2709	0.2676	0.2643	0.2611	0.2578	0.2546	0.2514	0.2483	0.2451
−0.7	0.242	0.2389	0.2358	0.2327	0.2296	0.2266	0.2236	0.2206	0.2177	0.2148
−0.8	0.2119	0.209	0.2061	0.2033	0.2005	0.1977	0.1949	0.1922	0.1894	0.1867
−0.9	0.1841	0.1814	0.1788	0.1762	0.1736	0.1711	0.1685	0.166	0.1635	0.1611
−1	0.1587	0.1562	0.1539	0.1515	0.1492	0.1469	0.1446	0.1423	0.1401	0.1379
−1.1	0.1357	0.1335	0.1314	0.1292	0.1271	0.1251	0.123	0.121	0.119	0.117
−1.2	0.1151	0.1131	0.1112	0.1093	0.1075	0.1056	0.1038	0.102	0.1003	0.0985
−1.3	0.0968	0.0951	0.0934	0.0918	0.0901	0.0885	0.0869	0.0853	0.0838	0.0823
−1.4	0.0808	0.0793	0.0778	0.0764	0.0749	0.0735	0.0721	0.0708	0.0694	0.0681
−1.5	0.0668	0.0655	0.0643	0.063	0.0618	0.0606	0.0594	0.0582	0.0571	0.0559
−1.6	0.0548	0.0537	0.0526	0.0516	0.0505	0.0495	0.0485	0.0475	0.0465	0.0455
−1.7	0.0446	0.0436	0.0427	0.0418	0.0409	0.0401	0.0392	0.0384	0.0375	0.0367
−1.8	0.0359	0.0351	0.0344	0.0336	0.0329	0.0322	0.0314	0.0307	0.0301	0.0294
−1.9	0.0287	0.0281	0.0274	0.0268	0.0262	0.0256	0.025	0.0244	0.0239	0.0233
−2	0.0228	0.0222	0.0217	0.0212	0.0207	0.0202	0.0197	0.0192	0.0188	0.0183
−2.1	0.0179	0.0174	0.017	0.0166	0.0162	0.0158	0.0154	0.015	0.0146	0.0143
−2.2	0.0139	0.0136	0.0132	0.0129	0.0125	0.0122	0.0119	0.0116	0.0113	0.011
−2.3	0.0107	0.0104	0.0102	0.0099	0.0096	0.0094	0.0091	0.0089	0.0087	0.0084
−2.4	0.0082	0.008	0.0078	0.0075	0.0073	0.0071	0.0069	0.0068	0.0066	0.0064
−2.5	0.0062	0.006	0.0059	0.0057	0.0055	0.0054	0.0052	0.0051	0.0049	0.0048

To obtain z find the nearest value of probability in the table; z is the addition of the number at the start of the row and the column. For example, for $p = 0.0244$, row $= -1.9$ and column $= -0.07$, hence $z = -1.9 - 0.07 = -1.97$.

Z	−0.06	−0.07	−0.08	−0.09
−1.8	0.0314	0.0307	0.0301	0.0294
−1.9	0.025	**0.0244**	0.0239	0.0233
−2	0.0197	0.0192	0.0188	0.0183
−2.1	0.0154	0.015	0.0146	0.0143
−2.2	0.0119	0.0116	0.0113	0.011
−2.3	0.0091	0.0089	0.0087	0.0084
−2.4	0.0069	0.0068	0.0066	0.0064
−2.5	0.0052	0.0051	0.0049	0.0048

ISO 14971 Annex C Pre–Risk Analysis Questionnaire

Questions	Applicable	Not Applicable	Comments
C.2.1 What is the intended use and how is the medical device to be used?			
2.1.1 What is the medical device's role relative to:			
2.2.1.1 Diagnosis, prevention, monitoring, treatment or alleviation of disease;			
2.2.1.2 Compensation for injury or handicap;			
2.2.1.3 Replacement or modification of anatomy, or control of conception?			
2.1.2 What are the indications for use (e.g. patient population)?			
2.1.3 Does the medical device sustain or support life?			
2.1.4 Is special intervention necessary in the case of failure of the medical device?			
C.2.2 Is the medical device intended to be implanted? Factors that should be considered include:			
2.2.1 The location of implantation;			
2.2.2 The characteristics of the patient population;			
2.2.3 Age;			
2.2.4 Weight;			
2.2.5 Physical activity;			
2.2.6 The effect of ageing on implant performance;			
2.2.7 The expected lifetime of the implant;			
2.2.8 The reversibility of the implantation.			
C.2.3 Is the medical device intended to be in contact with the patient or other persons? Factors that should be considered include the nature of the intended contact:			
2.3.1 Surface contact;			
2.3.2 Invasive contact;			
2.3.3 Implantation and, for each, the period and frequency of contact.			
C.2.4 What materials or components are utilized in the medical device or are used with, or are in contact with, the medical device? Factors that should be considered include:			
2.4.1 Compatibility with relevant substances;			
2.4.2 Compatibility with tissues or body fluids;			
2.4.3 Whether characteristics relevant to safety are known;			

Medical Device Design.
DOI: http://dx.doi.org/10.1016/B978-0-12-391942-7.00026-X

Questions	Applicable	Not Applicable	Comments
2.4.4 Is the device manufactured utilizing materials of animal origin? C.2.5 Is energy delivered to or extracted from the patient? Factors that should be considered include: 　　2.5.1 The type of energy transferred; 　　2.5.2 Its control, quality, quantity, intensity and duration; 　　2.5.3 Whether energy levels are higher than those currently used for similar devices. C.2.6 Are substances delivered to or extracted from the patient? Factors that should be considered include: 　　2.6.1 Whether the substance is delivered or extracted; 　　2.6.2 Whether it is a single substance or range of substances; 　　2.6.3 The maximum and minimum transfer rates and control thereof. C.2.7 Are biological materials processed by the medical device for subsequent reuse, transfusion or transplantation? Factors that should be considered include the type of process and substance(s) processed. C.2.8 Is the medical device supplied sterile or intended to be sterilized by the user, or are other microbiological controls applicable? Factors that should be considered include: 　　2.8.1 Whether the medical device is intended for single use or reuse packaging; 　　2.8.2 Shelf life issues; 　　2.8.3 Limitation on the number of reuse cycles; 　　2.8.4 Method of product sterilization; 　　2.8.5 The impact of other sterilization methods not intended by the manufacturer. C.2.9 Is the medical device intended to be routinely cleaned and disinfected by the user? 　　2.9.1 Factors that should be considered include the types of cleaning or disinfecting agents to be used and any limitations on the number of cleaning cycles. 　　2.9.2 The design of the medical device can influence the effectiveness of routine cleaning and disinfection. 　　2.9.3 In addition, consideration should be given to the effect of cleaning and disinfecting agents on the safety or performance of the device. C.2.10 Is the medical device intended to modify the patient environment? Factors that should be considered include: 　　2.10.1 Temperature; 　　2.10.2 Humidity; 　　2.10.3 Atmospheric gas composition; 　　2.10.4 Pressure; 　　2.10.5 Light;			

Questions	Applicable	Not Applicable	Comments

C.2.11 Are measurements taken?
    2.11.1 Factors that should be considered include the variables measured and the accuracy and the precision of the measurement results.
C.2.12 Is the medical device interpretative?
Factors that should be considered include:
    2.12.1 Whether conclusions are presented by the medical device from input or acquired data;
    2.12.2 The algorithms used;
    2.12.3 Confidence limits;
    2.12.4 Special attention should be given to unintended applications of the data or algorithm.
C.2.13 Is the medical device intended for use in conjunction with other medical devices, medicines or other medical technologies?
Factors that should be considered include:
    2.13.1 Identifying any other medical devices, medicines or other medical technologies that can be involved and the potential problems associated with such interactions, as well as patient compliance with the therapy.
C.2.14 Are there unwanted outputs of energy or substances?
Energy-related factors that should be considered include:
    2.14.1 Noise and vibration;
    2.14.2 Heat;
    2.14.3 Radiation (including ionizing, non-ionizing, and ultraviolet/visible/infrared radiation);
    2.14.4 Contact temperatures;
    2.14.5 Leakage currents;
    2.14.6 Electric or magnetic fields;
    2.14.7 Substance-related factors that should be considered include substances used in manufacturing;
    2.14.8 Cleaning or testing having unwanted physiological effects if they remain in the product;
    2.14.9 Other substance-related factors that should be considered include discharge of chemicals;
    2.14.10 Waste products;
    2.14.11 Body fluids.
C.2.15 Is the medical device susceptible to environmental influences?
Factors that should be considered include the operational, transport and storage environments. These include:
    2.15.1 Light;
    2.15.2 Temperature;
    2.15.3 Humidity;
    2.15.4 Vibrations;
    2.15.5 Spillage;
    2.15.6 Susceptibility to variations in power and cooling supplies;
    2.15.7 Electromagnetic interference.			

Questions	Applicable	Not Applicable	Comments

C.2.16 Does the medical device influence the environment? Factors that should be considered include:
     2.16.1 The effects on power and cooling supplies;
     2.16.2 Emission of toxic materials;
     2.16.3 The generation of electromagnetic disturbance.
 C.2.17 Are there essential consumables or accessories associated with the medical device?
     2.17.1 Factors that should be considered include specifications for such consumables or accessories and any restrictions placed upon users in their selection of these.
 C.2.18 Is maintenance or calibration necessary? Factors that should be considered include:
     2.18.1 Whether maintenance or calibration are to be carried out by the operator or user or by a specialist;
     2.18.1 Are special substances or equipment necessary for proper maintenance or calibration?
 C.2.19 Does the medical device contain software?
     2.19.1 Factors that should be considered include whether software is intended to be installed, verified, modified or exchanged by the operator or user or by a specialist.
 C.2.20 Does the medical device have a restricted shelf life?
     2.20.1 Factors that should be considered include labeling or indicators and the disposal of such medical devices when the expiration date is reached.
 C.2.21 Are there any delayed or long-term use effects? Factors that should be considered include ergonomic and cumulative effects:
     2.21.1 Pumps for saline that corrode over time;
     2.21.2 Mechanical fatigue;
     2.21.3 Loosening of straps and attachments;
     2.21.4 Vibration effects;
     2.21.5 Labels that wear or fall off;
     2.21.6 Long-term material degradation;
 C.2.22 To what mechanical forces will the medical device be subjected?
     2.22.1 Factors that should be considered include whether the forces to which the medical device will be subjected are under the control of the user or controlled by interaction with other persons.
 C.2.23 What determines the lifetime of the medical device?
     2.23.1 Factors that should be considered include ageing and battery depletion.
 C.2.24 Is the medical device intended for single use?
     2.24.1 Factors that should be considered include: Does the medical device self-destruct after use? Is it obvious that the device has been used?
 C.2.25 Is safe decommissioning or disposal of the medical device necessary?			

Questions	Applicable	Not Applicable	Comments
2.25.1 Factors that should be considered include the waste products that are generated during the disposal of the medical device itself. 2.25.2 Toxic or hazardous material; 2.25.3 Material recyclable? C.2.26 Does installation or use of the medical device require special training or special skills? 2.26.1 Factors that should be considered include the novelty of the medical device and the likely skill and training of the person installing the device. C.2.27 How will information for safe use be provided? Factors that should be considered include: 2.27.1 Whether information will be provided directly to the end-user by the manufacturer or will involve the participation of third parties such as installers, care providers, healthcare professionals or pharmacists and whether this will have implications for training; 2.27.2 Commissioning and handing over to the end-user and whether it is likely/possible that installation can be carried out by people without the necessary skills; 2.27.3 Based on the expected life of the device, whether re-training or re-certification of operators or service personnel would be required. C.2.28 Will new manufacturing processes need to be established or introduced? 2.28.1 Factors that should be considered include new technology or a new scale of production. C.2.29 Is successful application of the medical device critically dependent on human factors such as the user interface? C.2.29.1 Can the user interface design features contribute to use error? Factors that should be considered are user interface design features that can contribute to use error. Examples of interface design features include: C.2.29.2 Is the medical device used in an environment where distractions can cause use error? Factors that should be considered include: C.2.29.3 Does the medical device have connecting parts or accessories? Factors that should be considered include: C.2.29.4 Does the medical device have a control interface? Factors that should be considered include: C.2.29.5 Does the medical device display information? Factors that should be considered include: C.2.29.6 Is the medical device controlled by a menu? Factors that should be considered include: C.2.29.7 Will the medical device be used by persons with special needs?			

Questions	Applicable	Not Applicable	Comments

Factors that should be considered include:
    C.2.29.8 Can the user interface be used to initiate user actions?
Factors that should be considered include:
C.2.30 Does the medical device use an alarm system?
    2.30.1 Factors that should be considered are the risk of false alarms, missing alarms, disconnected alarm systems, unreliable remote alarm systems, and the medical staff's possibility of understanding how the alarm system works. Guidance for alarm systems is given in IEC 60601-1-8.
C.2.31 In what way(s) might the medical device be deliberately misused?
Factors that should be considered are:
    2.31.1 Incorrect use of connectors;
    2.31.2 Disabling safety features or alarms;
    2.31.3 Neglect of manufacturer's recommended maintenance.
C.2.32 Does the medical device hold data critical to patient care?
    2.32.1 Factors that should be considered include the consequence of the data being modified or corrupted.
C.2.33 Is the medical device intended to be mobile or portable?
Factors that should be considered are:
    2.33.1 The necessary grips;
    2.33.2 Handles;
    2.33.3 Wheels;
    2.33.4 Brakes;
    2.33.5 Mechanical stability and durability.
C.2.34 Does the use of the medical device depend on essential performance?
    2.34.1 Factors that should be considered are, for example, the characteristics of the output of life-supporting devices or the operation of an alarm. See IEC 60601-1 for a discussion of essential performance of medical electrical equipment and medical electrical systems.			

Generic Codes for Class I Medical Devices (MHRA)

Administration	
A1	Measuring and Mixing Devices for Medicines
A2	Inhalation Devices (e.g. Chamber Spacers)
A3	Sets – Solution/Irrigation (Gravity Only)
A4	Syringes (Hypodermic/Oral/Irrigation)
A5	Dispensers (Cement, etc.) and Accessories
A6	Sensitivity Testing Devices
A7	Non-Active Auto injector Devices
A8	Non-Active Infusion Devices and Accessories

Dental	
B1	Dental Lights
B2	Dental Diagnostic Fiber Optic Hand Pieces
B3	Dental Instruments (Reusable and Non-Powered)
B4	Dental Prophylaxis Paste (Non-Fluoride)
B5	Handheld Dental Mirrors and Accessories
B6	Impression Materials, Trays and Adhesives/Bite Wafers
B7	Orthodontic Materials (Extra-Oral/Intra-Oral Transient and Short-Term Use)
B8	Retraction Cords/Dental Wedges/Rubber Dam/Matrix Bands
B9	Articulating Paper/Spray
B10	Waxes
B11	Dental Unit Accessories
B12	Artificial Teeth
B13	Base Materials
B14	Dental Mouth Wash Tablets (Non-Medicated)
B15	Denture Lining Materials/Adhesives
Z169	Denture Cleaning Liquids/Tablets (Non-Disinfecting) (Dental Devices)
Z195	Dental Brace/Denture Fitting Aid (Dental Devices)
Z275	Denture Cleaning Brushes (Dental Devices)

Medical Device Design.
DOI: http://dx.doi.org/10.1016/B978-0-12-391942-7.00027-1

Diagnostic	
C1	Gels
C2	Electrodes/Transducers and Accessories
C3	Peak Flow Meters
C4	Sphygmomanometers and Accessories
C5	Stethoscopes
C6	Thermometers (Clinical)
C7	Examination/Procedure Gloves
C8	Blood Sampling Devices (Reusable)
C9	Endoscopes/Endoscopic Instruments and Accessories
C10	Image Storage and Retrieval System
C11	Laryngoscopes/Otoscopes and Accessories
C12	X-Ray Cassettes, Cassette Holders, Image Enhancers and Intensifying Screens
C13	Radiographic Film Processing Chemicals
C14	X-Ray Film Illuminators
C15	Sampling and Cell Collection Devices (Patient Contact – not IVDs)
Z102	Audiometer Accessories (Electro-Medical Mechanical Devices)
Z202	X-Ray Film Markers and Accessories (Diagnostic and Therapeutic Radiation Devices)
Z203	Patient Radiation Protection Products and Accessories (Diagnostic and Therapeutic Radiation Devices)

Dressings	
D1	Bandages (e.g. Support/Tubular/Adhesive/Plaster of Paris/Cast Liners/Resin)
D2	Cotton Wool/Gauze/Non Woven (Ribbons/Swab/Buds)
D3	Adhesive Plasters/Dressings/Tapes/Barrier Films
D4	Eye Occlusion Plasters/Shields and Corneal Shields
D5	Chiropody Dressings and Pads
O8	Wound Manager (Single Use)
Z149	Dressing Adhesive Removers (Single Use)

Equipment and Furnishing	
E1	Allergen Resistant Bedding
E2	Examination/Treatment Couches and Leg/Arm Rests
E3	Hospital Beds and Patient Positioning Aids
E4	Patient Hoists/Transfer Aids and Accessories
E5	Pressure Relief Devices and Accessories
E6	Treatment Chairs (Chiropody/Dental/Ophthalmic)
E7	Stretchers/Chairs/Hospital Trolleys (Patient Transport)
E8	Traction and Surgical Immobilization Devices
E9	Medical Examination Luminaries
E10	Rehabilitation Equipment
E11	Splints (Limb/Body/Ear)/Collars
E12	Resuscitation Devices (Non-Active) and Accessories

Equipment and Furnishing	
E13	Warming and Cooling Pads and Blankets (Non-Active and Non-Chemical)
O7	Speech/Breathing Training Devices (Technical Aids for Disabled Persons)
Z50	Cleaner/Washer for Medical Devices (Hospital Hardware)
Z125	Ultrasonic Cleaners and Solutions (Hospital Hardware)
Z131	Surgical Equipment Sterile Drapes (Hospital Hardware)
Z135	Autoclave Accessories (e.g. Trays and Tray Lifters, Shelves, Racks) (Hospital Hardware)
Z143	Speech Synthesizers/Communication Aids/Voice Amplification Systems (Technical Aids for Disabled Persons)
Z154	Tourniquets and Tourniquet Machines (Electro-Medical Mechanical Devices) (can use code O7 – Technical Aids for Disabled Persons pg 7)
Z170	Instrument Cleaning Solutions/Wipes (Non-Disinfecting)(Hospital Hardware)

Ophthalmic	
F1	Lamps (Ophthalmic Examination)
F2	Fundas Cameras/Keratometers/Slit Lamp Microscopes and Associated Software
F3	Low Vision Aids
F4	Operating Room Microscopes/Magnification Systems
F5	Ophthalmoscopes/Retinascopes
F6	Spectacle Lenses
F7	Spectacle Frames
F8	Ready-Made Spectacles (Non-Prescribed)
F9	Sight Testing Devices
O9	Schirmer Tear Test (Sterile Product) (Ophthalmic and Optical Devices)
Z45	Class I Tonometer (Reusable)
Z105	Eye Speculums (Ophthalmic and Optical Devices)
Z130	Contact Lens Accessories (Ophthalmic and Optical Devices)
Z148	Eye Baths/Irrigation Systems and Eyewash Solutions (Ophthalmic and Optical Devices)

Orthoses and Prostheses	
G1	Orthopedic Footwear
G2	Orthoses (Lower and Upper Limb/Spinal/Abdominal/Neck/Head)
G3	Trusses
G4	Compression Hosiery/Garments
G5	External Limb Prostheses and Accessories
G6	Stump Socks and Boards
G7	Orthopedic Casting/Support Products and Accessories
Z176	Postural Support Products (Technical Aids for Disabled Persons)

Surgical	
H1	Umbilical Clamps/Tape
H2	Tubes (Oesophageal/Rectal) and Accessories
H3	Enema and Douche Devices
H4	Incision Drapes/Theater Clothing
H5	Surgical Instruments (Reusable and Non-Powered)
H6	Pre-Operative Devices (Razor/Marker Pen)
H7	Airway Devices/Monitoring Equipment and Accessories
H8	Non-Invasive Drainage Devices and Accessories
H9	Surgical Instrument Accessories
H10	Sterilization Packaging
H11	Accessories for Implantable Devices (Non-Invasive)
H12	Operating Tables and Accessories
Z116	Vaginal Speculums (Reusable Devices)
Z136	Electrosurgical Accessories (e.g. Transient Invasive Electrodes, Footswitches, Electro-Medical Mechanical Devices)

Walking Aids and Wheelchairs	
I1	Crutch/Walking Stick
I2	Walking Frame/Multi-Leg Walking Aid/Standing Frame
I3	Rollator/Mobilator
I4	Wheelchairs (Non-Powered) and Accessories
I5	Wheelchairs (Powered) and Accessories
I6	Mobility Aids for the Visually Impaired
Z168	Rehabilitation Tricycles/Mobility Carts (Technical Aids for Disabled Persons)

Waste Collection	
J1	Ostomy Collection Devices and Accessories
J2	Incontinence Pads and Accessories
J3	Urinary Bags and Accessories
J4	Non-Invasive Tubing (Waste Disposal)
J5	Penile Sheath
J6	Urinary Catheters (Transient Use) and Accessories

Other	
Z48	Telemedicine Accessories (Reusable)
Z129	Acupressure Devices
Z146	Head Lice Devices (Reusable)
Z147	Dilators And Lubricants (Reusable)
Z162	Nasal Speculum (Reusable)
Z218	Lubricants (Instruments/Electrode Pads) (Single Use)

Custom-Made Devices	
K1	Dental Appliances/Prostheses
K2	Hearing Aid Inserts
K3	Prescribed Orthopedic Footwear
K4	Artificial Eyes
K5	Orthoses and Prostheses – External (Made Direct from Casts/Prescriptions)
K6	Orthopedic Implants
K7	Maxillo-Facial Devices
K8	Standing and Walking Frames
K9	Ligament and Tendon Repair Implants
K10	Spectacle Frames
Y4	Mandibular Advancement Device (Anesthetic and Respiratory Devices)
Y10	Insoles (Made Direct from Casts) (Technical Aids for Disabled Persons)
Y13	Postural Support Products (Technical Aids for Disabled Persons)
Y15	Splints (Limb/Body/Ear)/Collars (Single Use)

Procedure Packs	
L1	Ward Dressing Packs
L2	Theater Dressing Packs
L3	Oral Hygiene Packs
L4	First Aid Kits
L5	Prescribed Spectacles
L6	Cerebrospinal Fluid Filter With Syringe
L7	Ophthalmic Surgical Procedure Packs
L8	Orthodontic Procedure Packs
L9	Skin Traction Kits
L10	Surgical Procedure Packs (Includes Instruments Supplied Singularly)
X5	Theatre Drape Packs (Single Use)
X6	Needle Exchange Packs (Single Use)
X11	Endoscopes/Endoscopic Instruments and Accessories (Electro-Medical Mechanical Devices)
X14	Orthoses and Prostheses Procedure Packs (Single Use)
X18	Blood Specimen Collection Kits (Single Use)

FDA Class I and II Exempt Devices

Part 862 Clinical Chemistry and Clinical Toxicology Devices
Part 864 Hematology and Pathology Devices
Part 866 Immunology and Microbiology Devices
Part 868 Anesthesiology Devices
Part 870 Cardiovascular Devices
Part 872 Dental Devices
Part 874 Ear, Nose and Throat Devices
Part 876 Gastroenterology–Urology Devices

Part 878   General and Plastic Surgery Devices
Part 880   General Hospital and Personal Use Devices
Part 882   Neurological Devices
Part 884   Obstetrical and Gynecological Devices
Part 886   Ophthalmic Devices
Part 888   Orthopedic Devices
Part 890   Physical Medicine Devices
Part 892   Radiology Devices

Basic Materials Properties for Materials Selection

E.1 Density

Density is a measure of the amount of matter (mass M) compressed into an enclosed volume (V). Hence, more matter compressed into a space means *high* density, less matter means *low* density. It is commonly assumed that the density of a material is homogeneous (the same throughout the whole volume). Its symbol is, commonly, the Greek character rho, ρ; its units are normally in kg/m^3, though equivalent variants do exist.

For a specific body

$$density = \frac{mass}{volume}$$
$$\rho = \frac{M}{V} \left(\frac{kg}{m^3} \right)$$

(E.1)

Table E.1: Common Material Densities

Material	Density (kg/m³)	Lb/in³
Steels	7500–8080	0.271–0.292
Aluminum alloy	c3500	c0.126
Titanium	4500	0.163
Nylon	900–1120	0.0325–0.0405
PEEK	250–300	0.00903–0.0108

E.2 Stress and Strain

Designers are concerned about how materials fail. Consider a rod under tension by a force F (Figure E.1).

It is assumed that the force is evenly distributed across the whole area. This distribution is a ratio of the force (F) and the rod's cross-sectional area (A). This ratio is called *stress* and is given by the Greek character sigma, σ. Its units are commonly given as N/m^2 or Pa (both are the same).

Medical Device Design.
DOI: http://dx.doi.org/10.1016/B978-0-12-391942-7.00028-3

339

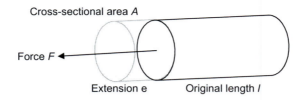

Figure E.1
Rod under stress.

$$stress = \frac{force}{cross\text{-}sectional\ area}$$
$$\sigma = \frac{F}{A}\left(\frac{N}{m^2}\right) \tag{E.2}$$

The rod will extend in the direction of the force applied. The measurement for this is called *strain*; its symbol is the Greek character epsilon, ε. It is defined as the ratio of the extension of the rod to the original length of the rod. It therefore has no units – it is a dimensionless quantity.

$$strain = \frac{extension}{original\ length}$$
$$\varepsilon = \frac{e}{l}\left(\frac{m}{m}\right) \tag{E.3}$$

Hooke found that these two properties were interlinked for all solid materials. Indeed he found that for all solids the relationship between stress and strain is linear; this is called Hooke's Law. If we plot a graph of stress versus strain, for any given material, the graph produced is very characteristic – see Figure E.2.

Figure E.2 illustrates an example stress–strain curve for a ductile material. The yield stress (σ_y) is the elastic limit of the material. If you deform a component to stress levels below this value the material will deform, but will return to the original shape after the load is released. If you load the component above this value the component will deform plastically, that is, some of the deformation will be permanent and your component has failed. All safety factors are related to the yield stress. A safety factor of 2, for example, means your maximum stress does not exceed $\sigma_y/2$.

For some materials the yield point is not easily identifiable. In this case a common estimate for yield stress is the proof stress. This is defined as the stress required to obtain a specific amount of permanent deformation. Often this is 0.1% permanent deformation and it is then called $\sigma_{0.01}$.

The yield stress and proof stress of a specific material is not constant. It is highly dependent on how the material is treated (e.g., how it is processed and/or heat treated).

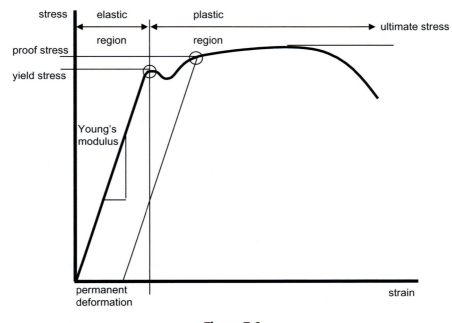

Figure E.2
Example graph of stress versus strain.

The flexibility of the material is related to Young's modulus. Its symbol is E and its units are also N/m^2. This is the slope of the stress-strain graph in the elastic region. It is relatively constant for a specific material. Materials with a high Young's modulus are stiff; those with low values are flexible (Table E.2).

CASE STUDY E.1

An orthopedic wire of diameter 2 mm is to be made from 316 L stainless steel annealed rod. If a safety factor of 2 is to be applied, determine the maximum tensile load.

Using matweb.com the material properties are

$$\sigma_y = 380\,\text{MN/m}^2$$

Using a safety factor of 2, this reduces to

$$\sigma_{max} = 380/2 = 190\,\text{MN/m}^2$$

Using Equation (E.2)

$$stress = \frac{force}{cross\text{-}sectional\ area}$$

$$\sigma = \frac{F}{A}$$

$$\therefore F = \sigma A$$

Hence

$$F = 190 \times 10^6 \times \pi r^2 = 190 \times 10^6 \times \pi 0.001^2 = 596.9\,\text{N}$$

Table E.2: Typical Material Properties for Some Common Materials

Material	σ_y (N/m^2)	E (GN/m^2)
Steels	110–2400	183–213
Aluminium alloy	1.24–750	c70
Titanium	170–310	110–120
Nylon	27–55	0.900–3.50
PEEK	90–140	2.70–12

Index